W9-AOP-686

FLORIDA STATE
UNIVERSITY LIBRARIES

APR 10 2000

TALLAHASSEE, FLORIDA

Active Tectonics and Alluvial Rivers

The interactions between tectonic uplift, river erosion and alluvial deposition are fundamental processes which have acted to shape the landscape we see today. These processes are of course ongoing, and are important not only in geomorphology, sedimentology and structural geology, but also hydrology and river engineering.

Stan Schumm, Jean Dumont and John Holbrook have combined their respective specialities to provide an overview of the effect of active tectonics on river morphology, behavior, and sedimentology. Their book brings together evidence and a variety of examples from both field and experimental studies to demonstrate how alluvial rivers are responding to uplift, subsidence and lateral tilting. Such recognition of the nature of river response yields criteria for the identification of active tectonics elsewhere, especially in areas without a history of seismic activity, or in the stratigraphic record. Using river characteristics such as sinuosity, gradient, and behavior to identify areas of active deformation can be of value in elucidating subsurface structure and in determining the cause of local flooding and problems of river stability of some of the world's great rivers, such as the Mississippi, Nile and Indus.

This volume will be of interest to graduate students, consultants and academic researchers in geomorphology, sedimentology and stratigraphy, structural geology, hydrology, geophysics, and geography.

Active
Tectonics
and Alluvial
Rivers

Stanley A. Schumm
Colorado State University

Jean F. Dumont
Laboratoire de Geodynamique sous-marine, ORSTOM

John M. Holbrook
Southeast Missouri State University

CAMBRIDGE
UNIVERSITY PRESS

GB
1203.2
.S36
2000

PUBLISHED BY THE PRESS SYNDICATE OF THE UNIVERSITY OF CAMBRIDGE
The Pitt Building, Trumpington Street, Cambridge, United Kingdom

CAMBRIDGE UNIVERSITY PRESS
The Edinburgh Building, Cambridge CB2 2RU, UK   http://www.cup.cam.ac.uk
40 West 20th Street, New York, NY 10011–4211, USA   http://www.cup.org
10 Stamford Road, Oakleigh, Melbourne 3166, Australia
Ruiz de Alarcón 13, 28014 Madrid, Spain

© Cambridge University Press 2000

This book is in copyright. Subject to statutory exception
and to the provisions of relevant collective licensing agreements,
no reproduction of any part may take place without
the written permission of Cambridge University Press.

First published 2000

Printed in the United Kingdom at the University Press, Cambridge

Typeface TEFFLexicon 10/14 pt   *System* QuarkXPress®   [SE]

A catalogue record for this book is available from the British Library

Library of Congress Cataloguing in Publication data
Schumm, Stanley Alfred, 1927–
Active tectonics and alluvial rivers/Stanley A. Schumm, Jean F. Dumont,
John M. Holbrook
      p.   cm.
ISBN 0 521 66110 2 (hb)
1. Alluvial streams.   2. Geology, Structural.
I. Dumont, Jean F. (Jean François), 1947–   .
II. Holbrook, John Millard, 1963–   .
III. Title.
GB1203.2.S36   2000
551.3′5 – dc21   99–24950 CIP

ISBN 0 521 66110 2 hardback

The feeling of colossal pleasure
lies chiefly in the consciousness
that something which you have
reckoned to be immovable,
has got it in it to move on its own.
That is probably one of the
strongest sensations of joy and
hope in the world. The dull globe,
the dead mass, the Earth itself rose
and stretched...

ISAK DINESEN
Out of Africa

Contents

Preface

The purpose of this book is to bring together evidence of the effect of active deformation on alluvial rivers. The emphasis will be upon morphologic adjustment, but where possible information on hydraulics, hydrology, and sedimentology is provided and applications to economic and structural geology, sedimentation and statigraphy, flood hydrology and river engineering are discussed.

The book is organized such that it has four parts:

> **Part I** consisting of Chapters 1 and 2 provides background information.
> **Part II** consisting of Chapters 3, 4 and 5 presents information on the effects of known measurable deformation.
> **Part III** consisting of Chapters 6 and 7 deals with the effects of less well documented (unmeasured) deformation on river characteristics.
> **Part IV** consisting of Chapters 8, 9 and 10 considers applications of the understanding of river response to deformation to sedimentology, geologic interpretation, and river engineering.

We believe that an effort such as this can be of value to river engineers, geomorphologists, economic and structural geologists, sedimentologists, and stratigraphers, and in our final chapter, we attempt to demonstrate this fact. We regret that some examples of river morphology are expressed in English units in Chapters 5 and 9. We considered converting to SI units, but because all river distances and locations on navigation charts and the location of navigation aids of the Mississippi are in English units, we elected to retain them. Also, if there is any interest in locating the sites described on maps or in the field, English units are needed.

Acknowledgments

A major part of this book is based upon unpublished reports prepared by Water Engineering & Technology and Resource Consultants & Engineers for the National Science Foundation and the U.S. Army Corps of Engineers. In addition, substantial portions of Chapter 3 were taken from the unpublished Ph.D. dissertation of Dr. Shunji Ouchi, and most of the material of Chapter 6 was taken from the unpublished Ph.D. dissertation of Dr. David Jorgensen. The senior author was significantly involved in all of the consulting reports and dissertation research and, therefore, the plagiarism is more apparent than real. We thank Ouchi and Jorgensen for permission to utilize their dissertations in this way.

Most of the data regarding subsiding foreland basins (part of Chapters 2 and 7) were obtained by ORSTOM (The French Institute of Scientific Research for Development in Cooperation) during research projects in South America. These projects were performed in cooperation with the IGP (Instituto Geofísico del Perú, in Lima, Peru), the IIAP (Instituto para la Investigación de la Amazonia Peruana, in Iquitos, Peru), the UMSA (Universidad Mayor San Andrés, La Paz, Bolivia) and the Semena (Trinidad, Bolivia).

Finally we are very grateful to Saundra Powell of Ayres Associates for her cheerful and efficient assistance in preparation of the manuscript.

Part I

Background

1

Introduction

Tectonics enters into every aspect of the earth sciences. Although it is possible to study erosional and depositional processes without concern for tectonics, the relief that permits these processes to continue is always in some way related to uplift or subsidence. For example, the erosional evolution of a landscape, as proposed by W.M. Davis, requires a tectonic boost to start the process (Davis, 1899; King and Schumm, 1980), and in an attempt to link tectonics to landforms, Walter Penck (1953) estimated relative rates of uplift and denudation from the shape of valley-side profiles.

Obviously, tectonics has had a role in earth history since the origin of the planet, and the interpretation of this history has been the traditional role of the structural geologist. More recently much attention has been paid to the role of tectonics in human affairs, and active tectonics has become a major concern with much emphasis on earthquake studies (McCalpin, 1996). Active tectonics is the ongoing deformation of the earth's surface (Wallace, 1985). More broadly, active tectonics is defined as "those tectonic processes that produce deformation of the earth's crust on a time scale of significance to human society" (Keller and Pinter, 1996, p. 2). As noted above, the major concern is with earthquake prediction. In order to predict and to understand active tectonics it must be studied within the context of the tectonic framework that probably developed during millions of years, which is the time span (post-Miocene) during which deformation is referred to as neotectonics. However, without measurements it is not possible to determine if one is dealing with ongoing active tectonics or geologically recent neotectonics. For us, the important factor is that the deformation is impacting a river, and it is the syntectonic response of the river that is of concern. Syntectonics refers to contemporaneous or coeval deformation and river response, which permits discussion of both active tectonic and neotectonic impacts on rivers. This is a largely ignored aspect of tectonic geomorphology, the study of landforms that result from tectonic processes (Yeats *et al.*, 1997, p. 139), which has been involved primarily with earthquake effects and prediction. Emphasis has been on the morphology and evolution of fault scarps, deformed river and marine terraces, and the morphology of mountain fronts (Bull, 1984; Morisawa and Hack, 1985; Merrits *et al.*, 1994; Keller and Printer, 1996; Keller *et al.*, 1998). We, on the other hand, are concerned with the effect of tectonics on the most susceptible of landforms, alluvial rivers and their deposits. The purpose is threefold as follows: to explain the variability of alluvial rivers, as affected by active deformation, to relate syntectonic river response to fluvial sedimentary deposits, flood hydrology, and hydraulics, and finally to consider briefly the practical significance of these findings for structural geologists, geomorphologists, sedimentologists, stratigraphers, economic geologists (petroleum), and river engineers.

Three books have been published that deal with structural landforms (Twidale, 1971; Tricart, 1974; Ollier, 1981), but the discussion of syntectonic effects on rivers is limited. In Tricart's (1974) book, the discussion centers on long-term effects of faulting and warping, the offsetting of river courses, the formation of lakes by faulting, and the effect of faulting on meanders, which may become very angular in plan.

Twidale (1971, pp. 133–6) recognizes the effect of faulting on drainage lines. He states that the rise of a fault block across a stream causes either the formation of a lake or swamp, or avulsion and the development of an irregular or abnormal drainage pattern. Twidale refers to the Murray River near Echuca in Victoria, Australia as a classic example of tectonic diversion caused by the rise of the Cadell Fault block (Figure 1.1), which has converted the Murray River from a single channel to two channels that flow around the obstruction (Bowler and Harford, 1966). The abandoned segment of the Murray River channel is preserved on the dipslope of the fault block.

Ollier (1981) devotes a chapter to drainage patterns, rivers and tectonics, and he discusses the effects of warping and faulting on drainage systems, but nothing on river morphology. However, recently Keller and Pinter (1996) have published a book on active tectonics, and they devote a chapter to rivers and drainage network response to deformation. More recently, Miall (1996) in his comprehensive work on fluvial deposits discussed the syndepositional effects of faults and folds.

Drainage patterns

Tectonic effects can be readily recognized in consolidated rocks, where stream channels and drainage networks have incised into and have

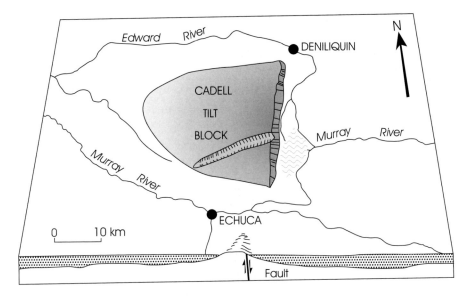

Figure 1.1 Disruption of Murray River, New South Wales, Australia, by Cadell Fault (modified after Bowler and Harford, 1966).

adjusted themselves to the varying resistance of rocks which compose the earth's surface. The best examples are the various types of drainage networks; for example, a rectangular drainage network forms as the result of intersecting joint sets or faults, and a trellis drainage network develops on folded strata (Figure 1.2, Table 1.1). In most cases, the effect of structure and tectonics is an accomplished fact. Nevertheless, there must have been a long period of adjustment, as the channels in the drainage networks responded to tectonic influences. If deformation was too rapid, undoubtedly there was a disruption of the existing river system. If deformation was slow, the existing river system could persist in its location. Deformation may not be continuous; it will probably be episodic, and it can cause earthquakes or it can be aseismic. An example of drainage network adjustment to tectonics is provided by Gupta (1997), who recognized the merging of several networks to form a gridiron or pitchfork pattern (Figure 1.3) in Nepal, as adjacent anticlines expanded laterally.

The landscape evolves as tectonically produced slopes are modified by erosion and deposition as well as by the continued growth of active structures (Figure 1.3). Drainage systems adapt to the changes of surface slope, and they have the potential to record information about the evolution of faults and folds (Ollier, 1981; Leeder and Jackson, 1993). For example, the drainage system on an anticline can mirror its structure and asymmetry (Figure 1.4A). In this case, there is a drainage divide situated close to the steeper flank. The

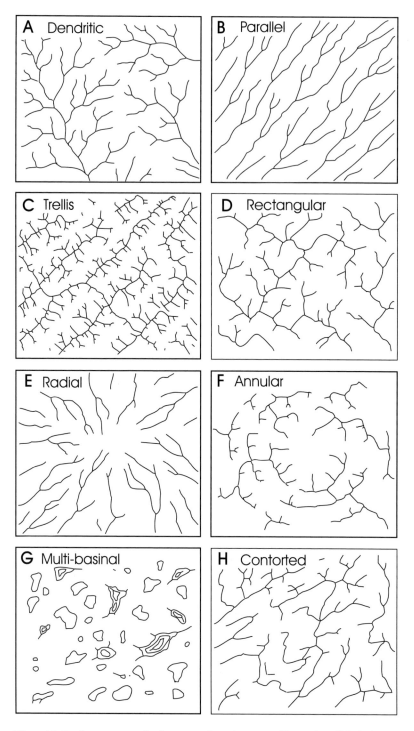

Figure 1.2 Drainage networks, for an explanation see Table 1.1 (modified after Howard, 1967).

Table 1.1. *Basic drainage patterns and their geologic significance. (see Figure 1.2)*

	Basic	Significance
A	Dendritic	Horizontal sediments or beveled, uniformly resistant, crystalline rocks. Gentle regional slope at present or at time of drainage inception. Type pattern resembles spreading oak or chestnut tree
B	Parallel	Generally indicates moderate to steep slopes but also found in areas of parallel, elongate landforms. All transitions possible between this pattern and type dendritic and trellis
C	Trellis	Dipping or folded sedimentary, volcanic, or low-grade metasedimentary rocks; areas of parallel fractures; exposed lake or sea floors ribbed by beach ridges. All transitions to parallel pattern. Type pattern is regarded here as one in which small tributaries are essentially same size on opposite sides of long parallel subsequent streams
D	Rectangular	Joints and/or faults at right angles. Lacks orderly repetitive quality of trellis pattern; streams and divides lack regional continuity
E	Radial	Volcanoes, domes, and erosion residuals. A complex of radial patterns in a volcanic field might be called multi-radial
F	Annular	Structural domes and basins, diatremes, and possibly stocks
G	Multi-basinal	Hummocky surficial deposits; differentially scoured or detailed bedrock; areas of recent volcanism, limestone solution, and permafrost. This descriptive term is suggested for all multiple-depression patterns whose exact origins are unknown
H	Contorted	Contorted, coarsely layered metamorphic rocks. Dikes, veins, and magmatized bands provide the resistant layers in some areas

Note:
From Howard (1967).

streams flow perpendicular to both the drainage divide and the structural contours on the uplifted surface. Although the structure of the Rock and Pillar range appears roughly symmetric (Figure 1.4B), its drainage system is very asymmetric. The drainage divide is located close to the front margin (southeast), across which there is a system of closely spaced streams flowing perpendicular to the divide and to the structural contours. The drainage pattern indicates that the symmetric box fold grew from an originally asymmetric fold related to movement of fault 1 (Figure 1.4B), which established the drainage divide along the southeast flank. The northwest flank was later elevated and steepened (fault 2), requiring the streams draining northwest to incise.

An important point to consider is that mountain ranges can be formed by

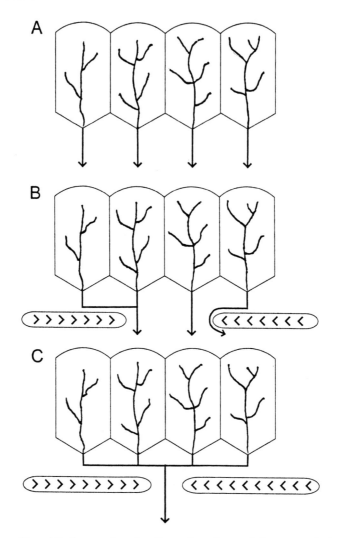

Figure 1.3 Cartoon showing disruption of several river networks by laterally growing anticlines to form a single network with a gridiron or pitchfork pattern (modified after Gupta, 1997).

the progressive longitudinal juxtaposition of discrete segments of the range. Jackson *et al.* (1996) analyzed longitudinal segment growth of the Blackstone and Raggedy Range and the Rough Ridge and South Rough Ridge, in the Alpine Range of New Zealand. The Blackstone and Raggedy Ranges constitute a continuous folded range schematically illustrated in Figure 1.5. The identification of the successive positions of the uplift is based on the occurrence of wind gaps on the drainage divide. Wind gaps on the Raggedy Range indicate that streams previously crossed the ridge south of the Ida Burn gorge (stage 1). This position has been progressively pushed northeast by the rising

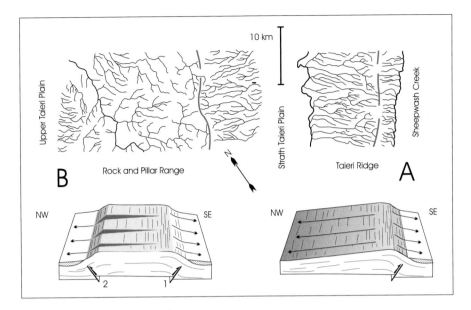

Figure 1.4 Maps and block diagrams of (A) Taieri Ridge and (B) Rock and Pillar Range, New Zealand (modified from Jackson *et al.*, 1996). Taieri Ridge has an asymmetric drainage pattern related to an asymmetric fold. The apparently symmetrical uplift of the Rock and Pillar Range has anomalous drainage and incised channels on one side, which indicates that the front thrust fault (1) occurred before the back thrust fault (2).

ridge of the Raggedy Range (stages 2 and 3). Increased width of the Raggedy Range uplift led to incision of Lauder Gorge.

Lateral growth of a folded range is also illustrated by the case of the Rough Ridge and South Rough Ridge near Oliverburn in Otago, New Zealand (Figure 1.6). Note the wind gaps on South Rough Ridge. They were formed by the drainage issuing from the front slope of Rough Ridge, prior to the uplift of the South Rough Ridge. The asymmetric pattern of the two major streams that cross South Rough Ridge suggests that the fault underlying it (2 in Figure 1.6) is later than the one forming Rough Ridge (1), and it has propagated to the north.

Jackson *et al.* (1996) propose an interesting check-list for the evaluation of drainage pattern development in segmented mountain ranges:

1. If the drainage divide (or fold axis) is perpendicular to the main streams on the flanks of the uplift (Figure 1.4), then the stream system is likely to be consequent with ridge development. If not, the stream system may be antecedent, predating the growth of the ridge.
2. If the asymmetry of the drainage is not mirrored in the asymmetry of

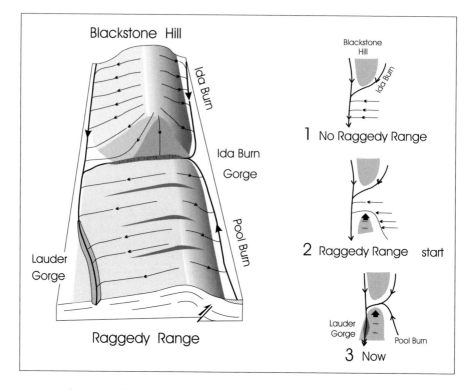

Figure 1.5 Schematic block diagram (left) illustrating the effect on drainage of propagation and joining of range segments, and interpretation of the evolution (right) (from Jackson *et al.*, 1996). Wind gaps were cut by previous drainage across the Raggedy Range (stage 1). As Raggedy Range was uplifted and extended to the north, the drainage was progressively directed toward the Ida Burn Gorge (stage 2). Water gaps are abandoned and Lauder Gorge was excavated as Raggedy Range uplift increased and the range expanded laterally (stage 3).

the topography (Figure 1.4), there may be other structural explanations.

3. The position of wind gaps on drainage divides may be significant for asserting both the relative ages (or activity) of structures and their directions of propagation (Figure 1.5).

4. Where antecedent streams cross ridges in gorges, the asymmetry of their catchment areas upstream may indicate the direction of propagation of the ridge (Figure 1.6).

5. Longitudinal stream courses may also contain clues to fault growth and interaction, particularly if they flow against the regional drainage trend, or incise through other structures (Figure 1.5).

The most interesting and challenging point of the Jackson *et al.* (1996) approach is the attempt to determine the age of deformation in relation to

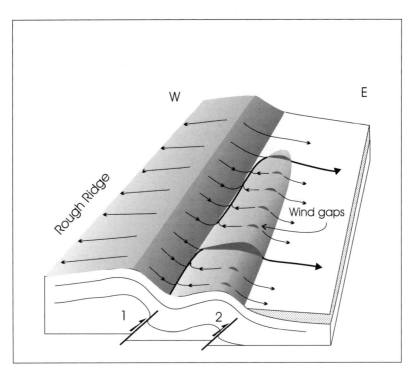

Figure 1.6 Schematic block diagram illustrating the structure of the drainage in a stepped range front (from Jackson *et al.*, 1996). The gathering of major streams into asymmetric catchments that cross the low front ridge suggest that fault 2 is later than fault 1. Water gaps formed as drainage from the front slope of Rough Ridge was superimposed during the early stage of uplift. They are preserved as wind gaps due to the increase of the uplift and the formation of a two-directional drainage on the flanks of the low front ridge

fault movement and seismic activity. Leeder and Jackson (1993) also examined the effects of surface deformation due to normal faulting on stream channel behavior. Seismotectonic considerations led them to conclude that drainage patterns can help to analyze the structure and evolution of fault segments. The vertical movements and tilting associated with normal faults can establish a drainage pattern that is strongly influenced by the continuity of the faults.

Alluvial rivers

Alluvial rivers flow through sediments that have been eroded and deposited by the river. That is, they are not significantly affected or constrained by bedrock or old terrace alluvium. Their morphology reflects a balance between the erosive power of the stream flow and the erosional

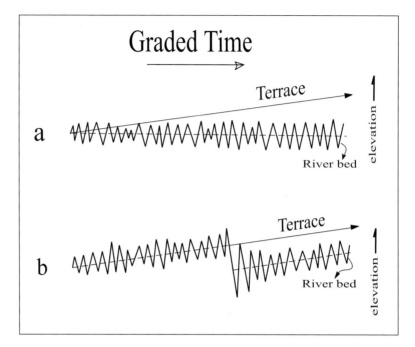

Figure 1.7 Schematic diagram showing two possible types of river response to active upwarping: (a) continuous adjustment, (b) adjustment which occurs when deformation exceeds a threshold.

resistance of the bed and bank materials. Changes in the morphology of an alluvial channel can occur as a result of changes in water discharge, sediment load, sediment type and gradient. Tectonic movement is only one way that the gradient of a river and its valley may be modified. Nevertheless, alluvial rivers can provide information concerning syntectonic deformation.

An alluvial river tends to attain an equilibrium (graded stream). The equilibrium is a condition in which any change in the controlling factors will cause a displacement of the equilibrium in a direction that will tend to absorb the effect of the change (Mackin, 1948). Slow uplift of the earth's surface may be accommodated by small essentially continuous incision (Figure 1.7a), or incision may not occur until the deformation becomes large enough for the river to exceed a threshold (Figure 1.7b). In the former case, slight progressive change will produce a persistent imbalance in river morphology, but in the latter case, the response should be abrupt. In any case, when an alluvial river responds and adjusts to deformation, a change of river morphology or channel behavior can be expected. Therefore, in order to understand and to predict changes in river morphology and channel behavior, it is necessary to know what effects crustal movement will have on alluvial rivers.

Active tectonics can take several forms. Deformation can be along faults or pairs of faults (horst and graben), which should have the same effect on rivers as a monocline, dome, or basin. In addition, an entire valley may be tilted upstream, downstream, or laterally. The possibilities are great, but, in reality, the primary effect of tectonics will be a local steepening or reduction of gradient or lateral (cross-valley) tilting. In addition to these primary influences, there will be secondary effects, as rivers respond to changed gradient (aggradation or degradation), and there will be tertiary effects, as decreased or increased sediment loads influence reaches downstream of the deformed reach and as aggradation or degradation in the deformed reach progress upstream.

Therefore, not only will the site of deformation (fault, fold) affect the channel at that location, but the effects can be propagated upstream and downstream. In addition, landslides triggered by seismic shock can introduce large quantities of sediment into a river, which may form temporary dams, and bank collapse into the river can greatly increase the sediment load and cause downstream channel adjustments.

Many of the major rivers of the world follow structural lows and major geofracture systems (Potter, 1978). Because of this structural control, some rivers have persisted in essentially their present location for as long as 1/16 of the earth's total history (Potter, 1978). In the United States, two major rivers that are clearly in areas of structural instability are the lower Mississippi River (Boyd and Schumm, 1995) and the Sacramento River (Fischer, 1994). Melton (1959) suspected that streams that have adjusted to tectonic activity are very numerous. He estimates that between 25 percent and 75 percent of all continental drainage in unglaciated regions has been tectonically influenced or controlled.

Wollard (1958) noted that a large number of earthquakes are associated with river valleys in the United States. His examples are the Columbia River where it swings north from a dominant east–west direction, the Mississippi River valley below the Ohio River junction, the lower reaches of the Wabash and Ohio Rivers and the Connecticut River valley. Elsewhere in the world, the upper and lower Amazon, Niger, Rhine, and Indus Rivers are located in areas of structural instability. The high discharge of the major rivers may permit them to maintain their courses in spite of local structural influences, and therefore, some of the clearest examples of the effect of tectonics on rivers may be found associated with smaller channels. However, the discharge of small channels is small, and their total energy is low; therefore, they may be unable to adjust rapidly to change. Large rivers, because of their low gradients, may in fact be the most significantly affected by minor changes in slope that are related to deformation. Little or nothing is known of these matters, and the literature on these topics is nonexistent.

Tectonic landforms

Deformation leading to river response is caused by many processes. Doming or subsidence of a region can result from large-scale isostatic rebound (Sparling, 1967), intrusion of salt domes (Collins *et al.*, 1981), differential compaction (Russell, 1939; Howard, 1967), solution (Benito *et al.*, 1997) subsurface ground-water and petroleum extraction (Yerkes and Castle, 1969; Prokopovich, 1983), fault movement in the subsurface (Russ, 1982), and pluton intrusion. Deformation of the bedrock floor of a valley is indicated by bedrock configuration and the thickness of overlying alluvium. For example, Kowalski and Radzikoska (1968) note that alluvium will be thickest over down-faulted blocks (graben) and thinnest over areas of uplift, as expected.

In spite of the practical significance of understanding active tectonics, relatively few investigators have considered its effects on alluvial rivers (Tator, 1958; Schumm, 1977, 1986; Adams, 1980; Russ, 1982; Burnett and Schumm, 1983; Dumont, 1994) and how these effects can be used to interpret tectonic activity. It is not surprising that little attention has been given to the effects of tectonics on alluvial rivers because variations of channel characteristics can also be attributed to downstream variations in discharge, sediment load, and the type of sediment moved through the channel (Schumm, 1977) or to local geology (Howard, 1967, p. 2256). Nevertheless, drainage anomalies that may be due to active tectonics have been identified and striking examples (Figure 1.8) involve lateral offset of streams along the San Andreas fault (Wallace, 1975; Sieh and Jahns, 1984). The large variety of types and rates of surficial movements and the difference among alluvial rivers add a further complication.

Usually geologists concentrate on the identification of geologic structures that are assumed to be quiescent (Howard, 1967; Ollier, 1981, p. 180; Deffontaines and Chorowicz, 1991) rather than on the effect of ongoing deformation. Nevertheless, DeBlieux (1949, 1962), DeBlieux and Shephard (1951), Tator (1958), and Lattman (1959) indicated that fluvial anomalies, such as local development of meanders or a braided pattern, local widening or narrowing of channels, anomalous ponds, marshes or alluvial fills, variations of levee width or discontinuous levees, and any anomalous curve or turn, are possible indicators of active tectonics (Figure 1.9), and these indicators have long been used in the search for petroleum.

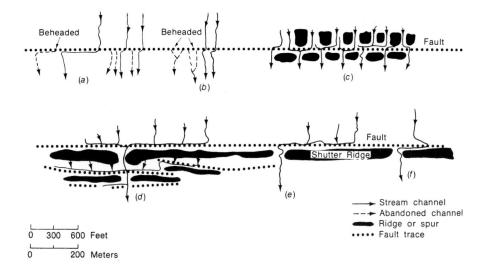

Figure 1.8 Patterns of fault-related stream channels found in the Carrizo Plain area, southern California (from Wallace, 1975). (a) Misalignment of single channels directly related to amount of fault displacement and age of channel. Beheading is common. (b) Paired stream channels misaligned. (c) Compound offsets of ridge spurs, and offset and deflection of channels. Both right and left deflection. (d) Trellis drainage produced by multiple fault strands, sliver ridges, and shutter ridges. (e) Offset plus deflection by shutter ridge may produce exaggerated apparent offset. (f) Capture by adjacent channel followed by right-lateral slip may produce "Z" pattern.

The investigation of active tectonics and its effects can be approached in two ways depending upon the objective of the investigation (Keller and Pinter, 1996, p. 52). Willemin and Knuepfer (1994) described these approaches as forward and inverse approaches. The forward approach involves the effects of known tectonic activity on landforms. The inverse approach involves interpreting the nature and location of active tectonics from anomalous landforms. The forward approach involves field and experimental studies of the effects of deformation on, in this book, alluvial rivers, whereas the inverse approach involves locating river anomalies that can be indicators of active tectonics. Examples of the first approach are presented in Chapters 3 and 4 and the second approach in Chapters 6–9. The second approach can lead to circular reasoning because anomalies that can be attributed to tectonics may be the result of other factors.

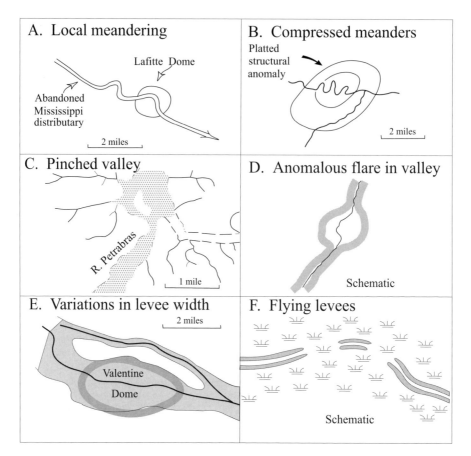

Figure 1.9 Examples of river and valley anomalies: A, E, F, Louisiana, (modified after DeBlieux, 1949). B, Kent County Texas (modified after DeBlieux and Shepherd, 1951). C, Amazon Basin (modified after Marple (1994). D, general example (modified after Howard, 1967).

Rates of deformation

Before proceeding, it is necessary to determine if rates of deformation are sufficiently rapid to cause modern channel adjustment. Obviously, surface disturbance associated with earthquakes (seismic deformation) occurs rapidly. For example, as a result of the 1964 Alaska earthquake, there was vertical deformation over an area of 170 000 to 200 000 square kilometers of land and sea-bottom in south-central Alaska. Uplift averaged about 2 m and the maximum uplift was about 12 m; elsewhere subsidence ranged from 1 to 2.3 m (Plafker, 1969).

Another type of deformation is aseismic, deformation not associated with earthquakes. This may be slow with progressive uplift or subsidence.

However, rates can be variable as in northern Israel (Kafri, 1969), where uplift velocities were much higher during the last few years of record than during the previous 20 years. Therefore, an oscillatory movement may be typical. Everything we know about geomorphology and tectonics indicates that the earth's surface does not behave in a regular manner, and the present rates of uplift in many areas are so rapid that they must cease and there must be a reversal. If not, there will be total disruption of drainage networks. For example, Schumm (1963) estimated that uplift at a rate of 8 m per 1000 years is not unusual. This is a rate of about 1 cm per year. Considering the flat gradients of many rivers, this rate of uplift could dramatically influence channel behavior.

Although the rate of active tectonic movements may vary widely, there seem to be limits. Kaizuka (1967) summarized worldwide rates of Quaternary vertical deformation, and he determined that the rates vary between 0.1 mm/year and 10 mm/year except for post-glacial rebound, which can be much greater. After the retreat of continental glaciers, wide areas in North America and Fennoscandia experienced rapid rates of rebound that have been declining with time (Farrand, 1962). The rate of uplift was high immediately following the removal of the excess weight of ice, but it becomes low when equilibrium is nearly established. Recent rates of uplift (from the late nineteenth century to the early twentieth century) in northwestern Europe, where the uplift is mainly the result of the post-glacial isostatic rebound, are estimated by Gutenberg (1941) to have a maximum of 10 mm/year and an average of 3 mm/year, but Andrews (1991) determined that average post-glacial uplift was 2 cm/yr.

Very rapid subsidence due to ground water withdrawal has been reported in many places in the world, including the United States (Poland, 1972, 1981). Subsidence is caused mainly by compaction of shallow unconsolidated clayey layers in response to pore-pressure reduction, which is induced by water table lowering. This rate of man-induced subsidence is much higher than tectonic movements. The maximum amount of subsidence reported in the United States exceeds 28 feet – in the San Joaquin Valley, California, for the period 1926 to 1969 (Poland et al., 1975). This is a rate of subsidence exceeding 200 mm/year, and maximum rates of subsidence exceeding 50 mm/year are not rare. A similar type of subsidence has been reported in some oil and gas fields (Yerkes and Castle, 1969).

A source of information from which evidence of active tectonic activity can be ascertained is the wealth of data compiled by the National Geodetic Survey (U.S. Dept. Commerce, 1972). For areas where repeated surveys have been made for an adequate period of time, they provide data of geological

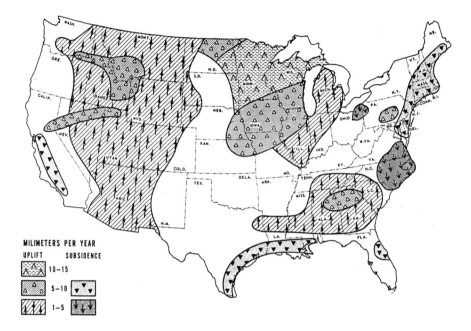

Figure 1.10 Vertical movements of the earth as determined from precise leveling by the National Geodetic Survey (U.S. Department of Commerce, 1972).

significance (Figure 1.10). These precise-leveling surveys can determine the vertical movement of a benchmark over a period of years, and a velocity in millimeters per year (mm/yr) can be computed. Along the survey line, the relative velocity of two benchmarks determines a tilt rate for the land surface along the survey route. Although Figure 1.10 is dated, it indicates that active tectonics occurs sufficiently rapidly to cause deformation of alluvial rivers, and such deformation can be expected almost anywhere. For example, the response of the earth's surface to deglaciation is variable, and it provides an indication of the flexibility of the geoid. Clark *et al.* (1978), identified six sea-level zones and each responded differently to deglaciation (Figure 1.11) as follows: Zone I is characterized by uplift due to isostatic adjustment, Zone II is a collapsing forebulge that fronted continental ice sheets, Zone III contains emerged beaches, Zone IV contains submerged coastlines, Zone V submerged but then minor emergence occurred, Zone VI is a coastal zone of uplift, as a result of meltwater loading of the continental shelves. Clearly, syntectonic river response can occur at many locations globally.

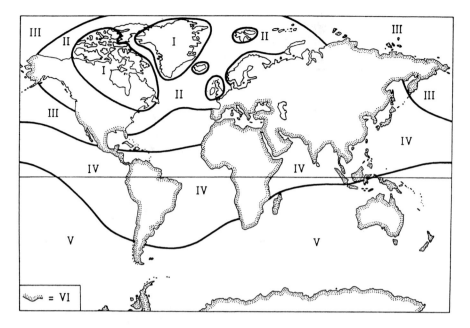

Figure 1.11 Six sea-level zones predicted by Clark *et al.* (1978) as a result of deglaciation during the last 16 000 years (see text for explanation).

Humans and active tectonics

Archaeologists have demonstrated that there was a long-term impact of active tectonics on ancient settlements in the Tigris and Euphrates Valley in Iraq and the Indus Valley of Pakistan. Ambraseys (1978) has been involved in seismo-tectonic studies throughout the Middle East and he believes that much surface deformation is aseismic in the Tigris and Euphrates Valley. In places, surface and underground canals have incised into their beds as a result of uplift. Elsewhere, erosion could not keep pace with the rise of the canal bed, and the water supply system was abandoned. Today locally the canal has a reverse gradient. Underground canals have been sheared off by faulting. Some of them have been repaired, but these new alignments are now offset by a few meters. For example, the Shaurn anticline in the southwestern corner of Iran forms a range of low hills that emerge from the alluvial plains between Shush and Ahwaz. Folding began in late Pliocene and it still continues. In the first or second century AD, two canals were cut across the anticline in order to lead water from a canal system on the northeast flank to the more extensive and fertile plains on the southwest. These canals afford a unique opportunity for measuring the uplift of the anticlines since about 1700 years ago. One canal still carries water, but where it crosses the anticline it has cut down about 4 meters below its original bed (2.4 mm/yr). The more northwesterly canal crosses a higher part of the anticline, and it was abandoned (Lees, 1955). Indeed, fluctuations in the water levels in the Los Angeles Aqueduct suggest deformation that may eventually cause water delivery problems in California (Leary *et al.*, 1981).

Another example of tectonic effects is in Iraq, where the character of the ancient irrigation systems associated with the Diyala River, a tributary of Tigris near Baghdad, has been altered by upwarping. Uplift, during the last thousand years, caused incision of the Diyala River into its alluvial plain and abandonment of irrigation canal systems (Adams, 1965). Abandonment of irrigated lands in Peru has also been attributed to uplift and canal incision (Moseley, 1983). On the other hand, uplift can cause backwater effects and sediment deposition, which creates new alluvial deposits that can be utilized for agriculture (Bailey *et al.*, 1993).

Discussion

Active tectonics and river response to it are topics of considerable concern for riparian landowners, river engineers, navigation interests, and the population in general whose taxes are required to maintain the channels or to compensate flood victims. Indeed, changes in river activity are important considerations in the siting of large industrial or commercial facilities or, in fact, any riparian structure. River changes may also provide evidence for the structural geologist of active tectonic features in areas of low relief and homogeneous lithology (Figure 1.9). In addition, syntectonic alluvial-river response will have an impact on fluvial sedimentary deposits, a major concern of sedimentologists and stratigraphers.

The investigation of active tectonics and its effects can be approached in two ways depending upon the objective of the investigation (Keller and Pinter, 1996, p. 52). Willemin and Knuepfer (1994) describe these as forward and inverse approaches. The forward approach involves the determination of the effects of known tectonic activity on landforms. The inverse approach involves the interpretation of the nature and location of active tectonics from a study of anomalous landforms. In this book, the forward approach involves field and experimental studies of the effects of known deformation on alluvial rivers, whereas the inverse approach involves locating river anomalies that can be indicators of active tectonics. Examples of the first approach are presented in Chapters 3 and 4 and the second approach in Chapters 6–9. As stated previously, care must be taken to avoid circular reasoning when the inverse approach is used.

2

Deformation and river response

Before specific examples of syntectonic impacts on alluvial rivers can be discussed, a brief review of types of tectonic activity, types of rivers, and river response will be presented. This background material sets the stage for detailed discussions that follow.

Types of deformation

The surficial movements in an alluvial valley can take different forms, as illustrated in Figure 2.1. The displacement can be seismic and associated with earthquakes and abrupt faulting, or it can be aseismic with progressive tilting and warping of the valley floor. Faults may be lateral faults that displace or offset the channel (Figure 2.1A) without vertical displacement. This type of displacement should be easily recognized. Faults with vertical displacement may have the uplifted block upstream of the fault with the result that gradient is steepened (Figure 2.1B). In the opposite case, the gradient will be decreased (Figure 2.1C). The effect will resemble monoclinal tilting (Figure 2.1F, G). Pairs of faults may produce uplifted (horst) or downdropped blocks (graben) that will both steepen and reduce gradient (Figure 2.1D, E). This has the same effect as domes and anticlines or basins and synclines (Figure 2.1H, I).

In addition to all of these structural features, the entire valley may be tilted upstream or downstream or the tilting may be across the valley toward either side of the floodplain (Figure 2.1J, K, L). The possibilities are great, but in reality the result will be local steepening or reduction of gradient or cross-valley tilting.

Streams respond to vertical displacement along faults (Figure 2.1B and C) by aggradation or degradation. When the displacement produces a channel segment steeper than the original stream gradient (Figure 2.1B), erosion will be initiated in this reach. When the displacement produces a channel segment

21

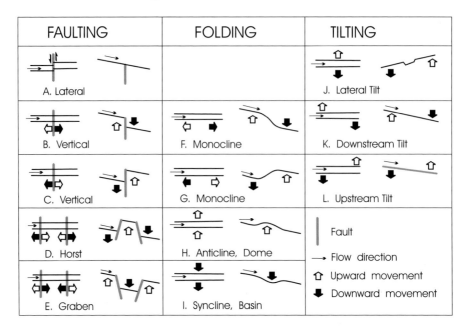

Figure 2.1 Surface deformation by faulting, folding, and lateral tilting. Plan view on left; cross-section on right (from Ouchi, 1983) small arrows indicate direction of flow. Large arrows indicate direction of movement. The displacement in each case is greatly exaggerated.

of lower elevation or gradient than the original stream (Figure 2.1C), aggradation will occur. Even small displacements may be sufficient to induce aggradation or degradation of large streams. If the displacement forms a dam (Figure 2.1C), the stream will be blocked, and it will flow along the fault or form a lake. Wallace (1968) mentioned that even a few inches of vertical upthrow along the downstream side of a fault can produce a dam across a small stream, which can divert its course.

Horst and graben (Figure 2.1D and E) combine two different types of vertical displacement (Figure 2.1B and C). There will be aggradation upstream from the horst, and degradation, which will migrate upstream, on the downstream part of the horst. Reduced sediment supply because of aggradation upstream will enhance the downstream degradation. Upstream of the graben, there will be degradation and aggradation in the graben itself. A river in such a location will be unstable, as Cartier and Alt (1982) suggested for the Bitterroot River in Montana.

The type of folding (Figure 2.1) will affect a river similar to the various types of faulting, but changes will probably be less abrupt. The effects of tilting will depend upon the amount, but in the simplest case, steepening of a valley will cause degradation (Figure 2.1K) and a reverse tilt (Figure 2.1L) will cause deposition. Lateral tilting will cause channel shift downdip (Figure 2.1J) or avulsion.

Figure 2.1 shows only simple, but exaggerated cases, whereas actual displacements can be more complicated. For example, faults can take any angle from parallel to perpendicular to river flow direction. Fault displacements of the surface at the time of an earthquake are obvious, and its influence on alluvial rivers can be observed. For example, small stream channels have been offset by strike-slip movement (Figures 1.8, 2.1A) of the San Andreas fault. Wallace (1968) pointed out that the offset of a stream channel depends on the relative rates of fluvial and tectonic processes. The main reason why channel offset caused by the San Andreas fault is clear is the extremely high rate of displacement along the fault (20.3 mm/year, Brown and Wallace, 1968) and the relatively small size of streams crossing and flowing along the fault. While fault displacements are obvious, movements of the land surface by folding can be slow. However, this type of deformation can also be an important influence on river behavior. For example, concentration of erosion along only one river bank can be the result of lateral tilt (Jefferson, 1907; Nanson, 1980a; Leeder and Alexander, 1987).

Types of alluvial rivers

It is apparent that different types of alluvial channels will respond differently to deformation; therefore, their characteristics must be reviewed before their response can be evaluated. This can best be done by discussing a simple classification of alluvial channels that is based on type of sediment load and pattern.

Five basic channel patterns exist (Figure 2.2): (1) straight channels with either migrating sand waves; or (2) with migrating alternate bars forming a sinuous thalweg; (3) two types of meandering channels, a highly sinuous channel of equal width (pattern 3a) and channels that are wider at bends than in crossings (pattern 3b); (4) the meandering–braided transition; and (5) a typical braided stream. The relative stability of these channels in terms of their normal erosional activity and the shape and gradient of the channels, as related to relative sediment size, load, velocity of flow, and stream power, are also indicated in Figure 2.2. It has been possible to develop these patterns experimentally by varying the gradient, sediment load, stream power, and the type of sediment load transported by the channel (Schumm and Kahn, 1972). Therefore, alluvial channels have also been classified according to the type of sediment load moving through the channels, as suspended-load, mixed-load,

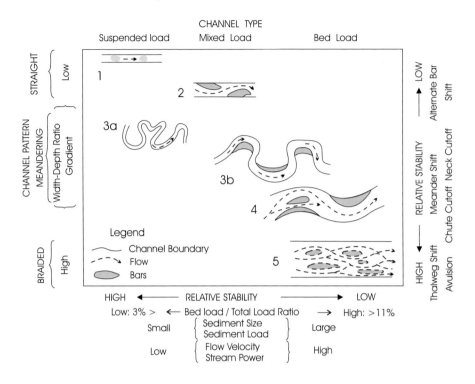

Figure 2.2 Channel classification based on pattern and type of sediment load with associated variables and relative stability indicated (Schumm, 1981).

and bed-load channels (Figure 2.2). Water discharge determines the dimensions of the channel (width, depth, meander dimensions), but the relative proportions of bed load (sand and gravel) and suspended load (silts and clays) determine not only the shape of the channel but width–depth ratio and channel pattern. A suspended-load channel has been defined as one that transports less than 3 percent bed load and a bed-load channel as one transporting more than 11 percent bed load (Schumm, 1977). The mixed-load channel lies between these two (Figure 2.2).

Alluvial rivers that transport clay, silt, sand, and gravel, can be placed within these three general categories. However, within the meandering-stream group there is considerable range of sinuosity (1.25 to 3.0). In addition, in the braided-stream category, there are bar-braided and island-braided channels (islands are vegetated bars).

Figure 2.2 suggests that the range of channels from straight to braided forms a continuum, but experimental work and field studies have indicated that the changes of pattern between braided, meandering, and straight, occur

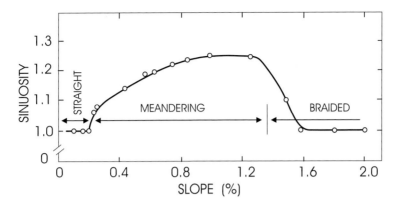

Figure 2.3 Relation between flume slope and sinuosity during experiments at constant water discharge. Sediment load, stream power, velocity increase with flume slope and a similar relation can be developed with these variables (from Schumm and Khan, 1972).

relatively abruptly at river-pattern thresholds (Figure 2.3). The pattern changes take place at critical values of stream power, gradient, and sediment load (Schumm and Kahn, 1972).

Although the five patterns of Figure 2.2 involve all three river types, there are five basic bed-load channel patterns (Figure 2.4A) that have been recognized during experimental studies of channel patterns (Schumm, 1977, p. 158). These five basic bed-load channel patterns can be extended to mixed-load and suspended-load channels to produce 13 river patterns (Figure 2.4). Patterns 1–5 are bed-load channel patterns, patterns 6–10 are mixed-load channel patterns, and patterns 11–13 are suspended-load channel patterns. Figure 2.4 attempts to show sinuosity differences and how the pattern thresholds change with increasing valley slope, stream power and sediment load for each channel type. The three major river types are controlled by type of sediment load, but within each type, the different patterns reflect increased valley slope, sediment load, and stream power.

The different bed-load channel patterns (Figure 2.4) can be described as follows: Pattern 1, straight, essentially equal-width channel with migrating sand waves. Pattern 2, alternate-bar channel with migrating side or alternate bars and a slightly sinuous thalweg; Pattern 3, low-sinuosity meandering channel with large alternate bars that develop chutes; Pattern 4, transitional meandering-thalweg braided channel. The large alternate bars or point bars have been dissected by chutes, but a meandering thalweg can be identified. Pattern 5 is a typical bar-braided channel.

As compared to the bed-load channels, the five mixed-load channels

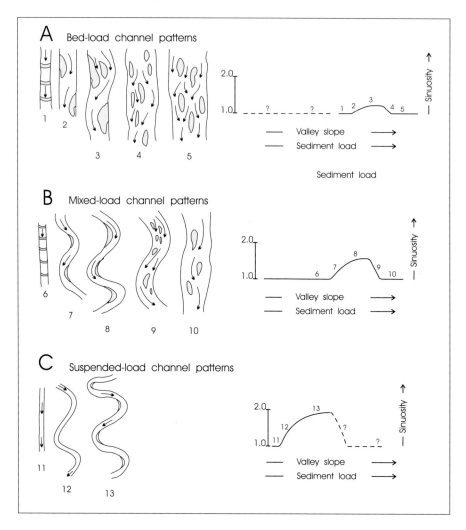

Figure 2.4 The range of alluvial channel patterns for the three channel types: (A) bed-load channel patterns, (B) mixed-load channel patterns, (C) suspended-load channel patterns (from Schumm, 1981).

(Figure 2.4B) are relatively narrower and deeper, and there is greater bank stability. The higher degree of bank stability permits the maintenance of narrow, deep, straight channels (Pattern 6), and alternate bars stabilize because of the finer sediments to form slightly sinuous channels (Pattern 7). Pattern 8 is a truly meandering channel, wide on the bends, relatively narrow at the crossings, and subject to chute cutoffs. Pattern 9 maintains the sinuosity of a meandering channel, but with a greater sediment transport the presence of bars gives it a composite sinuous-braided appearance. Pattern 10 is a braided channel that is relatively more stable than that of bed-load channel 5, and it is likely to be island-braided.

Suspended-load channels (Figure 2.4C) are narrow and deeper than mixed-load channels. Suspended load Pattern 11 is a straight, narrow, deep channel. With only small quantities of bed load, this type of channel may have the highest sinuosity of all (Patterns 12 and 13).

Rivers may undergo a metamorphosis during which the channel morphology changes completely. That is, a suspended-load channel (Pattern 12) could become braided (Pattern 5), or a braided channel (Pattern 5) could become meandering (Pattern 8 or 12), etc., when there is a sufficiently great change in the type of sediment load transported through that channel. Therefore, the change from one type of channel pattern to another may be relatively common, as the nature of the sediment moved through the system changes and this may be the effect of tributary sediment contributions, or upstream degradation or aggradation or tectonics.

There is an additional channel pattern that spans the range of stream types of Figure 2.4. This is the anastomosing pattern of branching and rejoining channel segments (Schumm *et al.*, 1996). The channel segments can be straight, meandering, or braided, and they appear to span the range of channel types. Anastomosing or anabranching channels differ from braided channels because they are composed of multiple channels that are separated by a floodplain (Figure 2.5), whereas braided channels have multiple thalwegs in a single channel. Knighton and Nanson (1993) and Richards *et al.* (1993) suggest that this pattern is transitioned to a single channel because anastomosing rivers appear to be associated with partly blocked valleys (Smith and Smith, 1980) and tectonic uplift (Gregory and Schumm, 1987). Reduced gradient from whatever cause appears to be important, and therefore, anastomosing reaches of a river can be evidence of tectonic activity.

Evidence of deformation

The first geomorphic clues of active tectonics may be tectonic landforms that result from long-continued deformation such as modified drainage networks and deformed sets of terraces. These obvious features are in contrast to the more subtle changes of alluvial rivers.

Drainage networks

The arrangement of streams and tributaries into a drainage network is affected by the regional slope of the surface on which the pattern develops, climate, and the erodibility of the surface material. Local variations in both

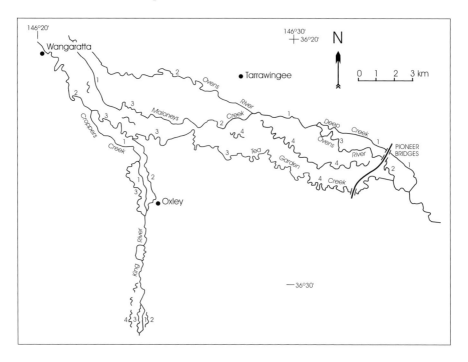

Figure 2.5 Map of anastomosing channels of Ovens and King Rivers. Numbers indicate relative ages of channels and reaches of channels from youngest (1) to oldest (4) (from Schumm *et al.*, 1996).

slope and materials will cause variations in the drainage pattern (Figure 1.2). For example, experimental studies demonstrated that parallel drainage patterns form on slopes greater than about 2.5 percent, whereas dendritic patterns form on gentler slopes (Phillips and Schumm, 1987).

Local variations in regional surface slope can cause anomalous drainage patterns. A topographic high caused by active uplift, will cause a deflection of portions of the drainage pattern. For instance, on the upslope side of an uplift the local surface slope will be opposite that of the regional slope. This can cause portions of the drainage pattern to develop in a direction opposite to that of the drainage network. An uplift may also deflect portions of the drainage from the general direction of the regional slope (Figure 2.6). As the major streams are deflected away from the uplift, tributaries on one side will be lengthened and those on the other side will be shortened.

Changes of an entire drainage network, as a result of tilting, are described by Sparling (1967) who notes that isostatic adjustment in Ottawa County, Ohio has increased the gradient of some streams and decreased the gradient of others. Incision of the steepened streams permitted them to capture the streams that were aggrading as a result of reduced gradient. Russell (1939)

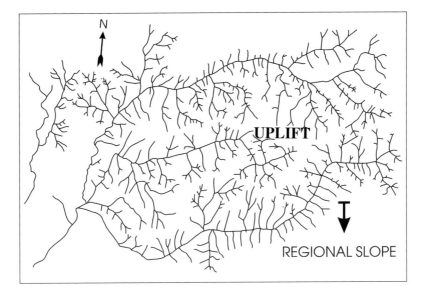

Figure 2.6 Drainage pattern modified by uplift.

described various drainage patterns found in the flat alluvial lands of Louisiana, and he recognized a network pattern of poorly developed drainage channels that was converted to a dendritic pattern, as a result of tilting and steepening of the gradient.

Price and Whetstone (1977) used asymmetric river and valley cross-section profiles, as evidence for movement along the Chattahoochee Embayment in Florida. Subsidence caused southward migration of east–west flowing channels. South-flowing streams exhibit paired terraces while east–west trending streams contain extensive terraces on the northern side with few or no terraces on the steeper southern margin. Subsidence in the Chattahoochee Embayment and adjacent warping along the Chattahoochee Uplift resulted in asymmetrical incision along the Chattahoochee River.

Long-continued tilting in a cross-valley direction will cause lateral erosion and the development of an asymmetrical valley and drainage network. Muehlberger (1979) noted this type of asymmetry in an area near Taos, New Mexico (Figure 2.7). Using 7½ minute U.S. Geological Survey topographic maps, he described this asymmetry quantitatively. By selecting a contour on the valley wall and measuring the distance from the stream to this contour in a down-tilt direction (Dd) and in the opposite, uptilt direction (Du) an index of asymmetry was developed (Dd/Du) (Figure 2.7). The index for 30 small drainage basins was 0.6 indicating a major displacement of the main channel in a down-tilt direction. Cox (1994) used a somewhat different index to

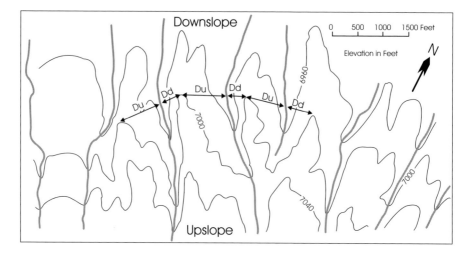

Figure 2.7 Map of asymmetrical stream valleys in the vicinity of Taos, New Mexico showing method of measuring distances used in calculating index of asymmetry (modified after Muehlberger, 1979).

demonstrate tilting and river shift in the southwestern Mississippi Embayment. Other indices of asymmetry are described by Keller and Pinter (1996, pp. 126, 127).

Lake patterns

Lakes are temporary features in terms of geological time, but they can be clear evidence of deformation. They occur at various scales, along primary or secondary portions of the drainage network, and generally disappear as erosion and deposition occurs through time. They may be isolated, but they are more frequently clustered in a specific area, or they are aligned along specific trends. Isolated lakes are generally due to local events such as landslides or collapses, which form a dam, but clustered lakes are due to regional processes, which may be related to climate (or paleoclimate), neotectonic deformation, or frequently a combination of these processes.

Large lakes may occupy geological basins, and active tectonics is partly or totally involved in their formation. Such lakes act as drainage collectors and they can occupy various elevations. The Bolivian Altiplano lakes (Titicaca, Popoo, Uyuni) stand at an elevation of around 4000 m and are associated with plate convergence and uplift. Continental stretching and transtensional tectonics are favorable for the formation of basins and tectonic lakes. Often the relation of lakes to faults is obvious, and a study of the lake pattern is not necessary to demonstrate the effect of tectonics. (See Chapter 5 where

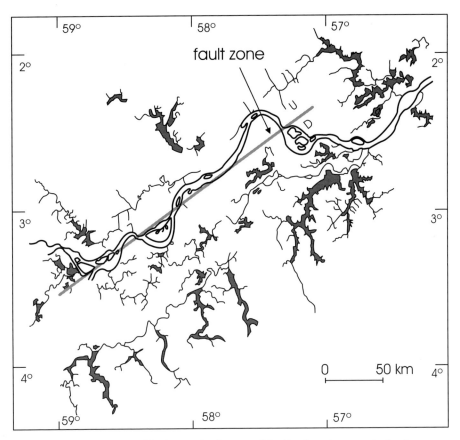

Figure 2.8 Inferred fault zone in the area of the Tupinambaranas, in the Amazon Basin between Obidos and Manaus Brazil (modified after Sternberg, 1955). The river trace and ria lakes suggest a downward movement along the same axis as that of the inferred fault.

Reelfoot Lake and the Mississippi valley sunklands are described; Figures 5.1, 5.2).

This discussion deals with cases where specific lake patterns, (ria lakes and elongated lakes), are related to active deformation. Most of these concepts are not new. For example, the significance of ria lakes was identified more than 30 years ago.

Ria Lakes

Dendritic ria lakes occur at the lower part of a tributary, which is blocked by aggradation along the trunk river (Holz *et al.*, 1979). In the lower and central parts of the Amazonian Basin the formation of ria lakes is related to post-glacial eustacy (Irion, 1984). Some of these lakes have a "knee" pattern superposed on the dendritic pattern suggesting the effect of active faulting through a thick sedimentary cover (Sternberg, 1950, 1955, Figure 2.8).

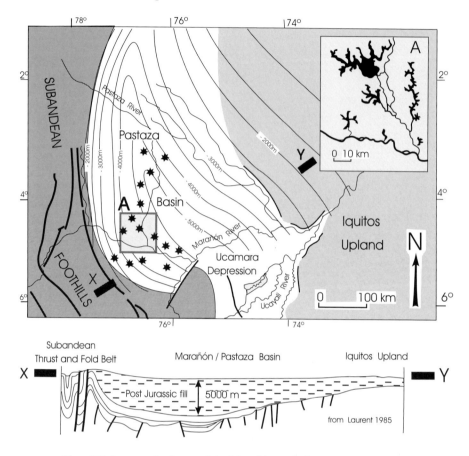

Figure 2.9 Structural scheme of the Marañón Basin in Peru, composed of the Pastaza Basin and the Ucamara Depression. Stars show the location of the main ria lakes, with the area A represented in detail. The contour lines show the isobath of the pre-Jurassic basement, from Sanz (1974). Section along X–Y in the lower part of the figure, modified after Laurent (1985).

Reservoirs behind dams take the form of ria lakes. Clusters of ria lakes are observed along large rivers in the western Marañón Basin and in the central Ucayali Basin, Peru (Räsänen *et al.*, 1987; Dumont, 1993). They are related to active subsidence of the back side of piggy back or thrust faults. Similar patterns are observed on the south margin of the Beni Basin (Dumont, 1992; Dumont and Guyot, 1993).

An elongated cluster of ria lakes is located along the lower Pastaza River and the Marañón River in the western part of the Marañón Basin (Figure 2.9). This cluster is superimposed over the structural axis of the basin, which trends toward the Ucamara Depression. This axis of maximum subsidence of the basin is located in front of the Subandean Thrust and Fault Belt (STFB).

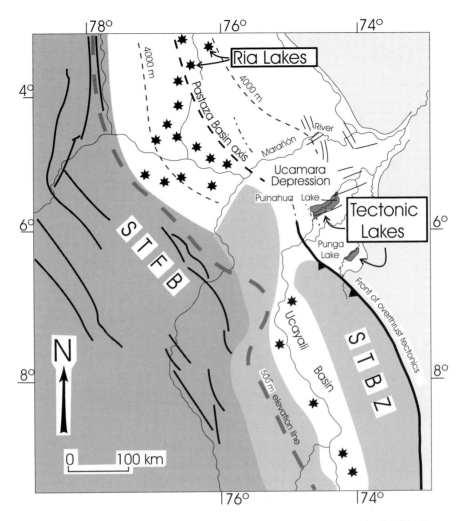

Figure 2.10 Relative positions of tectonic lakes in the Peruvian foreland basins. Stars indicate location of clusters of ria lakes (modified after Dumont , 1996).

The Pastaza depression is filled by more than 4500 m of Cretaceous and Cenozoic sediments, with about 500 m of Quaternary sediment (Laurent, 1985) (Figure 2.9). No ria lakes are observed in the south part of the Marañón Basin (Ucamara Depression), but ria lakes occur again in the Ucayali Basin of Central Peru (Figure 2.10), a piggy-back basin.

A continental rift generates tilt of the side blocks in opposite directions and the progressive tilting of the earth's surface can significantly modify the gradient of rivers. For example, Lake Victoria is drained by the Victoria Nile which flows north to Lake Kyoga (Figure 2.11). Lake Kyoga and Lake Kwania look artificial, but they are not the result of dam construction. They are, in fact,

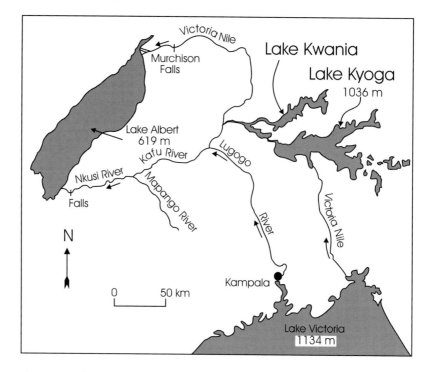

Figure 2.11 Lake Kyoga region, Uganda. Arrows show flow directions in rivers. Back tilting explains the shape of Lake Kyoga and Lake Kwania (modified after Doornkamp and Temple, 1966).

formed by uplift and eastward tilting of western Uganda. Flow in the Kafu River has been reversed, and water draining from Lake Victoria has found a new course to the north, where it flows over Murchison Falls and into Lake Albert. The geologically recent derangement of these drainage systems is the result of uplift that is apparently continuing at the present time (Doornkamp and Temple, 1966).

Elongated lakes

The term elongated lakes is applied to lakes that are long and narrow. These lakes are superimposed over basement structures. In subsiding basins, they are also closely related to active tectonics.

The Puinahua (1324 km²) and Punga (341 km²) lakes are both located in the south part of the Marañón Basin (Figure 2.10). They are long and narrow and wider on the foothill side than on the craton side. Punga Lake is superimposed over the structure of the Santa Elena uplift (Dumont and Garcia, 1991), which is interpreted as a crystalline horst surrounded by Paleozoic sedimentary strata (Laurent, 1985). The NE elongation of the lake is parallel to a few of the structural features mentioned by Laurent (1985). Both lakes are also parallel to the strike of normal faulting with a NNW–SSE extension in the upland border (Dumont *et al.*, 1988). As a result, the lakes are interpreted as the surface expression of tensional stress superimposed over reactivated basement structures due to the onset of Andean tectonics (Dumont and Garcia, 1991).

More precise morphological evidence of active tectonics comes from the Punga Lake. According to testimony of old settlers and published travel journals (Stiglish, 1904) the area was covered by forest, that developed on a terrace before 1923. Then the region began to subside. The tree trunks of the drowned forest are still visible, now below 2 m of water during low water stages. In this very flat area the water level rises about 2 m during floods, suggesting a minimum subsidence of 4 m over about 70 years. A tectonic interpretation is favored because a local rise of base level cannot be involved. Only a downward motion of the area where the lake is observed can explain the phenomenon. These lakes are related to downdropped blocks that are related to a tensional system.

Terraces

The vertical warping of terraces, as a result of deformation, has been studied by numerous investigators (Machida, 1960; Zuchiewicz, 1980; King and Stein, 1983). The offset of terraces by lateral faulting gives a clear indication of fault movement. The offset terraces at the mouth of the Waiohine Gorge in New Zealand shows the amount of displacement and the episodic nature of the displacement (Figure 2.12). In addition to river terraces, the deformation of lake terraces and marine terraces (Keller and Pinter, 1996) are proof of isostatic and tectonic activity because they formed horizontally at a given water level. An example is the isostatic deformation of the Pleistocene Lake Bonneville shorelines, Utah (Crittenden, 1963). Portions of the Lake Bonneville shorelines have been deformed as much as 210 feet, as it drained and evaporated to form Great Salt Lake. In addition, deformed marine terraces provide an excellent indication of recent deformation along 440 km of the Pacific Coast of Baja, California (Orme, 1980). Not only are terraces deformed, but as an uplift is crossed, a floodplain can be converted to a low terrace. This will, of course, dramatically alter the hydraulics and hydrology of the reach.

In addition to the obvious evidence of deformation described above, evidence of deformation is provided by alluvial deposits and paleosoils (Machette, 1978; Keller *et al.*, 1982; Bull, 1984; Rockwell *et al.*, 1984; Rockwell, 1988) by variations in gradients of streams of different orders (Merritts and

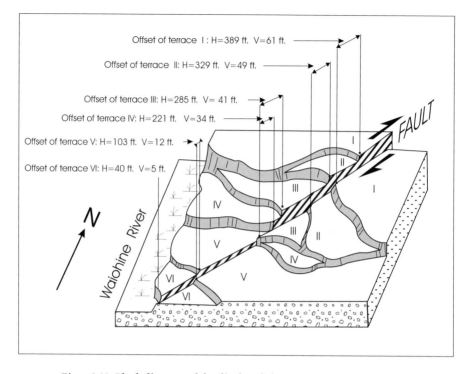

Figure 2.12 Block diagram of the displaced river terraces at the mouth of the Waiohine Gorge, New Zealand. The West Wairarapa Fault extends from bottom left (southwest) to top right (northeast) and has cut and moved successive river terraces (I = oldest terrace; VI = youngest terrace). The amount (in feet) each terrace has been moved by the fault is indicated (H = amount of horizontal movement; V = amount of vertical movement). The length of the fault shown on the diagram is about 0.8 km (1/2 mile) (modified after Stevens, 1974).

Vincent, 1989; Merritts and Hesterberg, 1994) and by variations of valley and river longitudinal profiles (Volkov *et al.*, 1967; Seeber and Gornitz, 1983).

River response

The clearest evidence for tectonic effects on rivers are anomalous reaches that show dramatic changes of pattern trend and gradient that cannot be attributed to other causes. Structural geologists and petroleum geologists have known for years that rivers are strong indicators of faulting (Lattman, 1959; Howard, 1967).

Tectonic activity can significantly control river patterns and behavior, and this is especially true of alluvial rivers. Neef (1966) and Radulescu (1962) state that neotectonic movements can be reflected only in those geomorphic features that react to the smallest changes of slope, such as meander characteristics. In addition, the variations of thickness and distribution of recent sediments indicate the variability of tectonics in a given valley. Clearly, one of the most sensitive indicators of change is the valley floor profile and longitudinal profile of a stream (Bendefy *et al.*, 1967; Zuchiewicz, 1979).

Alluvial rivers will be very sensitive indicators of valley slope change. In order to maintain a constant gradient, a river that is being steepened by a downstream tilt will increase its sinuosity or braid, whereas a reduction of valley slope will lead to a reduction of sinuosity or aggradation if the pattern cannot change. For example, Twidale (1966) reported that both the Flinders and Leichardt Rivers have changed to a braided pattern, as a result of the steepening of their gradient by the Selwyn Upwarp in northern Queensland, Australia; and the Red River near Winnipeg, Canada is straightening, as a result of reduction of gradient by isostatic rebound (Welch, 1973). Other examples are provided by the Mississippi River between St. Louis and Cairo and the lower Missouri River (Adams, 1980) as well as the Red River of the North near Winnipeg, Canada (Vanicek and Nagy, 1980).

Baselevel changes can also resemble the effects of active tectonics, and can provide information on channel change. For example, if sea-level is lowered, and a steep portion of the continental shelf is exposed, the effect will be like that downstream of the axis of an uplift. Lane (1955) argued that a lowering of baselevel will cause channel adjustment throughout most of the river. Indeed, Lane's (1955) argument that a stream will restore its gradient at a higher or lower level following a baselevel change is conceptually correct because if the river is initially at grade, then to move its sediment load and water discharge through the channel, the original gradient of the stream must be reestablished. However, sediment loads may be greater after channel incision and less after aggradation. Another fallacy in Lane's argument is the assumption that the valley slope and channel gradient are identical. This is not the case, and the two-dimensional perspective leads to erroneous conclusions. For example, a sinuous river has a gradient less than that of the valley floor.

Sinuosity (P) is the ratio between channel length (L_c) and valley length (L_v), and it is also the ratio between valley slope (S_v) and channel gradient (S_c) as follows:

$$P = \frac{L_c}{L_v} = \frac{S_v}{S_c} \tag{1}$$

Therefore, a straight channel has a sinuosity of 1.0, and the gradient of the channel and the slope of the valley floor are the same. It has been demonstrated that rivers can respond to major changes of water and sediment load primarily by pattern changes (Schumm, 1968), and that much of the pattern variability of large alluvial rivers such as the Mississippi, Indus, and Nile,

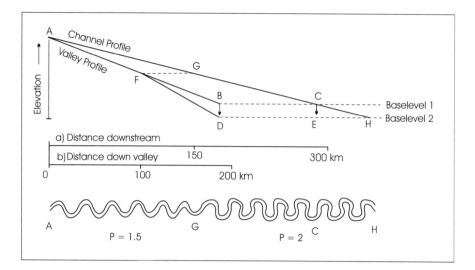

Figure 2.13 Effect of a baselevel fall on channel length and pattern. See text for discussion.

reflect the variability of the valley slope (Schumm *et al.*, 1994; Jorgensen *et al.*, 1993; Schumm and Galay, 1994).

Figure 2.13 illustrates this concept geometrically, and it shows the impact of the lowering of baselevel in a valley with a stream of sinuosity (*P*) 1.5. The line A–C represents the channel profile, and the line A–B represents the profile of the valley floor. Points B and C are at the river mouth, and points F and G are at the same location in the valley. The channel distance is one-third longer than the valley distance, and the difference in channel and valley slope reflects the sinuosity of the stream. The length of the channel is 1.5 times the length of the valley and, therefore, the stream gradient is one-third less than the valley slope (equation 1). If a vertical fall of baselevel from B to D and C to E is assumed, channel incision and lateral erosion will steepen the valley floor. If the channel is not confined laterally, it can adjust to the increased valley slope (F–D) by increasing sinuosity to 2.0, and the channel profile is extended to H. In this case, incision ceases at point F in the valley and at point G in the channel because the increase of sinuosity from 1.5 to 2.0 from G to H maintains a constant channel gradient over the reach of increased valley slope (F–D). The one-third increase of channel length (sinuosity) between G and H compensates for a one-third steepening of the valley floor from F to D.

According to Lane's assumptions, the effect of this baselevel fall would be propagated upstream to point A, where an amount of erosion equal to B–D would occur. However, because the stream can adjust, the steepening of the valley floor will not result in a change of stream gradient. Rather the channel lengthens, and the effect of baselevel lowering is propagated only a relatively short distance upstream. The distance will undoubtedly depend on local conditions and the original slope of the valley floor, but this exercise supports Saucier's (1991) contention that Pleistocene sea level change in the lower Mississippi valley was effective only as far as Baton Rouge (see also Blum, 1993 and Blum and Valastro, 1994). The probability that a large river can adjust in this fashion is made more likely by the fact that the baselevel changes in nature will take place relatively slowly and not abruptly, unlike during the experimental studies. The river, therefore, has more time to adjust by changing sinuosity.

Further evidence for the type of channel response shown in Figure 2.13 is demonstrated by the experimental studies of Jeff Ware (1992 oral comm.). He lowered baselevel relatively slowly to a maximum of 12 cm in a flume with a total length of 18.4 m. This change would have doubled the channel gradient. However, the effect of the baselevel lowering extended only 4 m upstream, and the change in baselevel was accommodated by an increase of sinuosity from 1.2 to 1.5 in the lower 4 m in the flume. It is clear, however, that pattern change did not totally compensate for the change of baselevel. This channel widened and roughness increased, thereby assuming part of the adjustment to the baselevel change.

Ware's experiments showed that a sinuosity increase, which resulted in a slope decrease, was only part of the adjustment, and width, depth, and roughness adjusted to decrease velocity and stream power. Therefore, channel pattern change may only absorb part of a valley slope or baselevel change. The River Nile provides an example of such shared adjustment with both sinuosity and width changing as valley slope changes (Schumm and Galay, 1994).

Changes of valley-floor gradient provide an explanation of downstream pattern variations, but variations of valley-floor slope can be the result of several influences. Tectonic activity may change the slope of the valley floor and have its effect on the channel pattern (Adams, 1980). In addition, a high-sediment-transporting tributary may build a fan-like deposit in the valley, which will persist even after the tributary sediment load has decreased. When the main river crosses this fan, pattern changes will result, as the river attempts to maintain a constant gradient. Tributaries to the Jordan River in Israel have developed fan-like deposits in the valley, and the valley floor of the Jordan Valley undulates as a result. The Jordan River, as it approaches one of these convexities, straightens as it crosses the upstream flatter part of the fan and then it develops a more sinuous course on the steeper downstream side of the fan (Schumm, 1977, p. 140).

It is important to realize that channels that lie near a pattern threshold

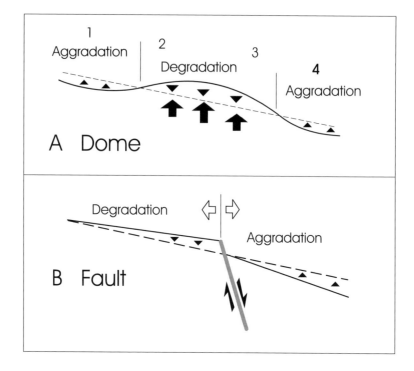

Figure 2.14 Reaches of degradation and aggradation associated with uplift and faulting.

(Figure 2.3) may change their characteristics dramatically with only a slight change in the controlling variable. For example, some rivers that are meandering and that are near pattern thresholds become braided with only a small addition of bed load (Schumm, 1979). Experimental studies and field observations confirm that a change of valley-floor slope will cause a change of channel morphology. The change will differ, however, depending where the channel lies on a plot such as that of Figure 2.3 and depending on the type of channel (Figure 2.4). For example, with increasing slope, a straight channel will become sinuous, a low sinuosity channel will become more sinuous, and a meandering channel will braid. With decreasing slope, a braided channel will meander and a meandering channel will straighten.

Although pattern changes may dominate river response, deposition in reaches of reduced gradient is likely as is channel incision and bank erosion in reaches of steepened gradient (Figure 2.14). In fact, upstream deposition will reduce downstream sediment loads thereby increasing the tendency for downstream erosion. Therefore, in addition to the primary valley-floor deformation by active tectonics and the secondary channel response to this deformation, there are third-order effects beyond the area of deformation. For

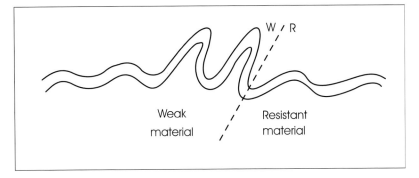

Figure 2.15 Effect of change from weaker (W) to more resistant (R) materials on meander pattern.

example, deposition upstream from the axis of a dome can progress further upstream by backfilling beyond the area of active deformation (Figure 2.14). This means that the uplift is acting as a dam, and unless erosion on the steeper downstream side of the uplift produces sufficient sediment to compensate, erosion will occur downstream beyond the limits of deformation. However, it is more likely that reduced sediment load will accelerate erosion on the steeper downstream reach, and when this increased load moves downstream, aggradation will result (Figure 2.14).

Another aspect of both active and neotectonics is the appearance of more resistant materials in the channel as the channel degrades. Resistant alluvium (clay plugs, gravel armor) or bedrock will confine the channel and retard meander shift and bank erosion. The result should be deformed or compressed meanders upstream and a change of meander character at the contact. For example, as meanders shift downvalley, their movement may be retarded if more resistant alluvium or bedrock is encountered (Jin and Schumm, 1987). Hence, a fault may present a barrier with the result that upstream meanders are compressed and deformed (Figure 2.15). A similar pattern will result if the river encounters bedrock as it crosses an upwarp or if, as a result of incision, it encounters resistant materials in a portion of its course. Yeromenko and Ivanov (1977) reviewed the Russian literature and concluded that for rivers crossing uplifts, the largest number of meanders occur upstream of the structures, which supports Gardner's (1975) observations of bedrock channels in the Colorado Plateau. This is the reverse of what is expected for alluvial channels from the relation of Figure 2.3. Therefore, an investigator must not simply base conclusions on river pattern alone.

Discussion

Alluvium is usually assumed to hide the underlying geology, but rates of active tectonic movement are sufficiently rapid to affect the morphology and behavior of alluvial rivers. Rivers being the most sensitive components of the landscape will provide evidence of even, slow, aseismic tectonic activity. For example, evidence of active tectonics is as follows:

1. deformation of valley floor longitudinal profile
2. deformation of channel longitudinal profile
3. change of channel pattern
4. change of channel width and depth
5. conversion of floodplain to a low terrace
6. reaches of active channel incision or lateral shift
7. effects both upstream and downstream of the zone of deformation (degradation, aggradation, flooding, bank erosion)
8. formation of lakes

Alluvial channels are sensitive indicators of change, but they also adjust to changes of hydrology and sediment load as well as to active tectonics. Therefore, it may be difficult to determine the cause of channel change when, in fact, human activities have been changing both discharge and sediment load during historic time. Pattern change alone is not sufficient evidence for active tectonics, rather it is one bit of evidence that must be supported with other morphologic evidence and/or surveys that provide clear evidence of deformation. In many areas, the evidence will be circumstantial. Nevertheless, anomalous reaches that are not related to artificial controls, tributary influences, lithologic change, or paleotectonics may reasonably be assumed to be the result of active tectonics until proved otherwise.

Lakes can also provide information on subsidence, and ria lakes can be evidence of tilting or even hydrologic and climatic influences. For example, during Pleistocene aggradation in the Ohio River and Mississippi River valleys, numerous lakes formed in tributary valleys. Therefore, care must be exercised when determining the tectonic significance of lakes and other river anomalies.

Part II

Effects of known measurable deformation

3

Experimental studies

The previous chapter was a discussion of syntectonic effects on alluvial rivers in general. According to Figure 2.3, tectonic movement changes valley slope and thus stream power, thereby producing a change in channel pattern. Hence, deformation of the valley floor by active tectonics can influence alluvial rivers. Experimental studies were undertaken to investigate the effect of valley (flume) slope deformation on channel planform characteristics. The advantages of the flume are that conditions can be controlled in order to investigate variations in a single variable, and the time scale is reduced by the rapid change which takes place in the flume (Schumm *et al.*, 1987).

Experimental studies by Ouchi (1983, 1985), Jin (1983), Dykstra (1988), and Peakall (1996), were designed to determine the effects of uplift, subsidence, and lateral tilt on rivers. Ouchi and Jin used a 9.1 m long, 2.4 m wide and 0.6 m high wooden flume (Figure 3.1). The central part of the flume had a flexible bottom, which could be moved up and down by two hydraulic jacks. Germanowski's (Germanowski and Schumm, 1993) study of aggrading and degrading braided streams in a 1.8 m wide and 18.3 m long flume also provides information on braided stream response to deformation.

Braided-channel response

Ouchi's (1983) experimental results are summarized in Figure 3.2. The change of channel pattern was not great for the braided channels (Figure 3.3) except where degradation occurred on the axis of the uplift and upstream and downstream of the zone of subsidence. With uplift, the reduction of slope caused deposition upstream of the uplift axis and deposition downstream, as the channel incised through the uplift (Figure 3.3A). With subsidence, the channel incised on the steepened upstream reach, and the trapping of sediment in the zone of subsidence caused degradation downstream (Figure 3.3).

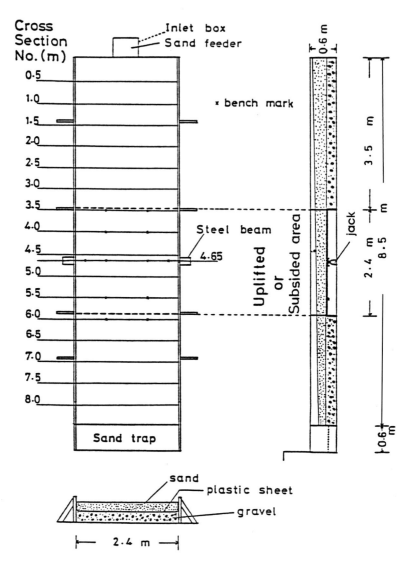

Figure 3.1 Diagram of flume used in modeling experiments (from Ouchi, 1983).

The slope changes affected the hydrology of the channel with higher water levels and aggradation causing bar submergence and thalweg shift upstream of the uplift axis and overbank flooding in the zone of reduced gradient during subsidence (Figure 3.2).

Experimental studies that were undertaken to study the effect of aggradation and degradation on braided-stream patterns can also be used to indicate the effects of active tectonics on braided river channels (Germanoski and Schumm, 1993). In a large flume, Germanoski performed two sets of

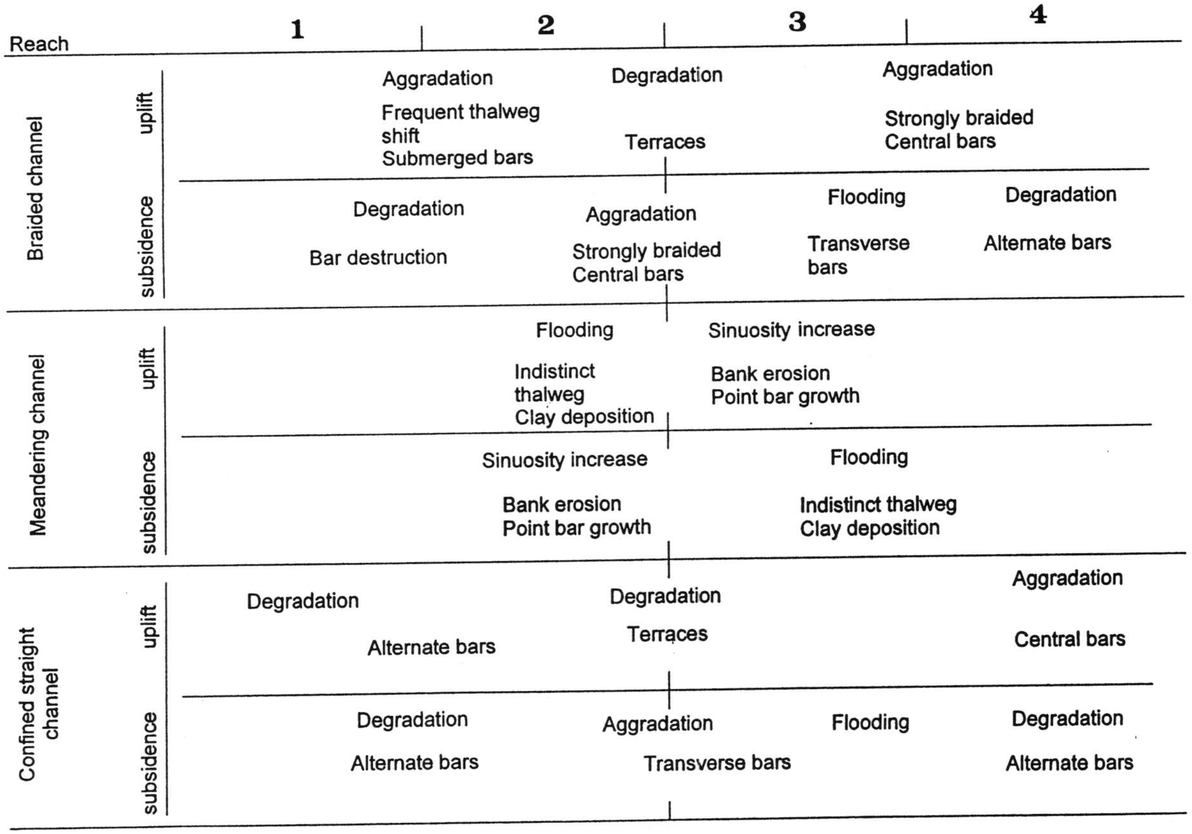

Reach

	1	2	3	4
Braided channel — uplift	Aggradation / Frequent thalweg shift / Submerged bars	Degradation / Terraces	Aggradation	Strongly braided / Central bars
Braided channel — subsidence	Degradation / Bar destruction	Aggradation / Strongly braided / Central bars	Flooding / Transverse bars	Degradation / Alternate bars
Meandering channel — uplift		Flooding / Indistinct thalweg / Clay deposition	Sinuosity increase / Bank erosion / Point bar growth	
Meandering channel — subsidence	Sinuosity increase / Bank erosion / Point bar growth		Flooding / Indistinct thalweg / Clay deposition	
Confined straight channel — uplift	Degradation / Alternate bars	Degradation / Terraces	Aggradation / Central bars	
Confined straight channel — subsidence	Degradation / Alternate bars	Aggradation / Transverse bars	Flooding	Degradation / Alternate bars

Figure 3.2 Responses of experimental channels to uplift and to subsidence for braided, meandering and straight channels. Reaches 2 and 3 are zones of deformation. Reach 4 is downstream, and Reach 1 is upstream of the zone of deformation.

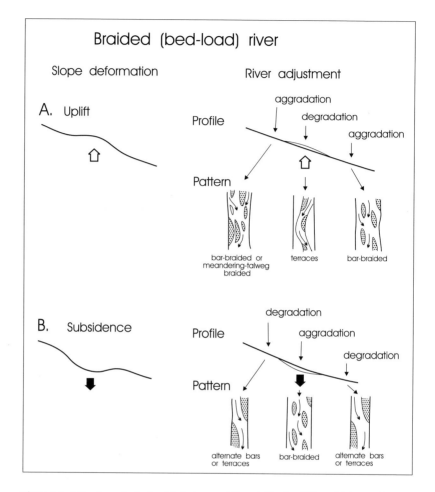

Figure 3.3 Adjustment of a braided river to (A) anticlinal uplift and (B) synclinal subsidence (from Ouchi, 1985).

experiments using medium sand (d_{50} = 0.87 mm) and fine gravel (d_{50} = 2.2 mm). Aggradation and degradation were induced by increasing or decreasing the sediment fed into the channel. Variations of channel patterns were described by number of bars, size of bars, number of linguoid dunes, and a braiding index. The braiding index (B.I.) was calculated as follows:

$$\text{B.I.} = \left[\frac{2(\text{Total length of bars})}{\text{length of reach}} \right] + \left[\frac{\text{Total number of bars}}{\text{length of reach}} \right]$$

For both the sand (Figure 3.4) and gravel channels (Figure 3.5) aggradation increased the number of braid bars (Figure 3.6) and increased the braiding index (Figure 3.7). Aggradation did not affect braid-bar size (Figure 3.8). It

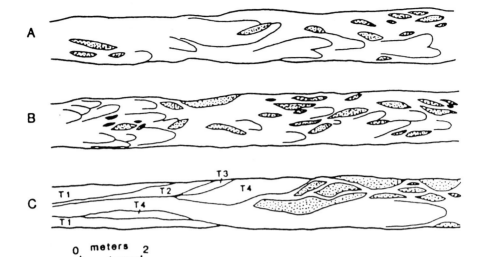

Figure 3.4 Maps of (A) equilibrium, (B) aggrading, and (C) degrading sand-bed channels. Aggradation (B) is reflected by an increase in the number of braid bars and an increase in braiding index, and degradation (C) is reflected by a decrease in braiding index, the development of a fewer number of larger braid bars formed from the coalescence of smaller bars in the downstream reach, and the development of an incised single-channel reach flanked by terraces in the upstream section of the channel (terraces are numbered from oldest T1 to youngest T4) (from Germanoski and Schumm, 1993).

increased the number of linguoid dunes in the sand channel (Figure 3.9) but not in the gravel channel. The change from equilibrium to degradation decreased the number of braid bars in the sand channels (Figure 3.6) and braid-bar size increased (Figure 3.8). There was no clear change of the other variables, although the braiding index for sand channels decreased in all cases (Figure 3.7). All of this suggests that if there is a significant change of braided river morphology, it could indicate that the channel is aggrading or degrading, as a result of tectonic influences. However, aggradation and degradation can be caused by other factors as well.

These morphological adjustments are the same as those reported by Maizels (1979) based upon her study of aggradation on the outwash train of the Bossons Glacier in the French Alps. As the Bossons valley train aggraded between 1968 and 1974, the braiding intensity and pattern complexity increased, the number of braid bars increased, braid-bar size decreased, and active braid-plain width increased (Maizels, 1979). The similarity between the laboratory channels and the Bossons outwash system suggests that the laboratory results are valid.

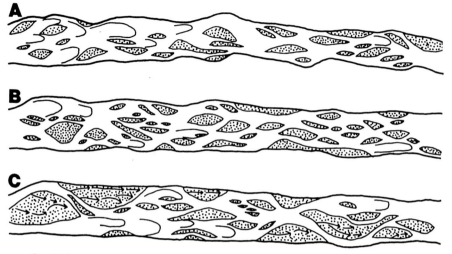

Figure 3.5 Maps of (A) equilibrium, (B) aggrading, and (C) degrading gravel-bed channels. Aggradation (B) is reflected by an increase in the number of braid bars present and an increase in braiding index, and degradation (C) is reflected by a decrease in braiding index and the development of fewer large braid bars formed from the coalescence of smaller bars. Dashed lines with arrow heads in channel C represent abandoned cross-bar channels (from Germanowski and Schumm, 1993).

Meandering-channel response

In the case of anticlinal or domal uplift across a meandering river, Ouchi (1985) concluded that there will be a sinuosity increase on the downstream side of the uplift as the valley floor is steepened (Figures 3.2, 3.10 and 3.11). On the upstream side of the uplift channel, straightening can be expected, but the damming effects of the uplift may be more apparent. Flood stage will increase as a result, and there will be inundation of the floodplain and channel avulsion. Judging from the experimental results, a swampy condition with deposition of fine sediments will occur. An anastomosing channel pattern may develop (Figures 3.10 and 3.11). Meander cutoffs on the steepened slope downstream will aid degradation, and temporarily a braided pattern may develop on the downstream side of the uplift in a mixed-load river (Figure 3.10), and cutoffs will straighten the suspended-load river (Figure 3.11) until it can re-establish a characteristic sinuous pattern.

A sinuosity increase also occurs on the upstream side of subsidence (Figures 3.12 and 3.13). In the downstream part of the subsidence, a condition similar to that occurring on the upstream side of the uplift is expected to occur.

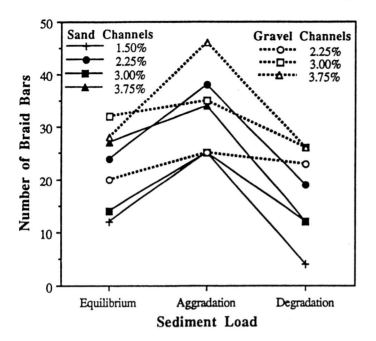

Figure 3.6 Relations between the number of braid bars present and sediment load and channel gradient in sand-bed channels and gravel-bed channels. Lines connecting data points on these and subsequent figures are not meant to imply linear trends. Percentage values in the figure "keys" are flume gradients (from Germanowski and Schumm, 1993).

However, because slope adjustment by aggradation is a slow process, a swampy condition in the lower downstream part of the subsidence can develop. The anastomosing pattern is most likely to be formed near the downstream end of subsidence for meandering rivers. A braided pattern for a mixed-load river (Figure 3.12) and the anastomosing pattern for a suspended-load river (Figure 3.13) should develop on the upstream side of the subsidence. Widespread anastomosing channels or lakes are likely to be common features in the area of subsidence.

Additional experiments to determine the effect of uplift on a meandering river were carried out by Jin (Jin and Schumm, 1987). Unlike Ouchi's experiments, which were all in sand, Jin constructed a floodplain that provided greater stability to the experimental channel. A mixture of kaolinite and fine sand (2 cm thick) was placed over a medium-sized sand base. The fine upper layer simulated less erosive floodplain sediments. Water was introduced into a straight channel, which after 400 hours became meandering (Figure 3.14).

After 400 hours, the center of the flume (Figure 3.1) was uplifted with hydraulic jacks to simulate tectonic uplift. The experiment was continued for

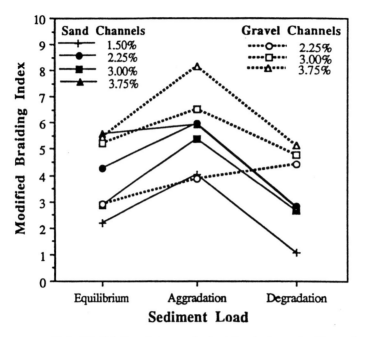

Figure 3.7 Relations between the braiding index and sediment load and channel gradient in sand-bed channels and gravel-bed channels. Percentage values in the figure "keys" are flume gradients (from Germanowski and Schumm, 1993).

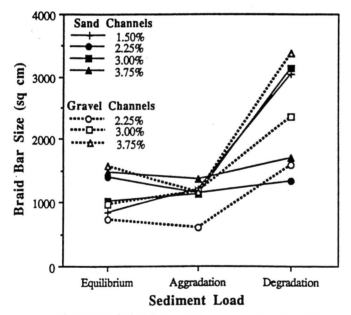

Figure 3.8 Relations between the average size of braid bars and sediment load and channel gradient in sand-bed channels and gravel-bed channels. Percentage values in the figure "keys" are flume gradients (from Germanowski and Schumm, 1993).

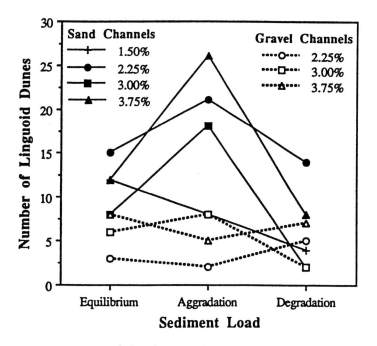

Figure 3.9 Relations between the number of linguoid dunes and sediment load and channel gradient in sand-bed channels and gravel-bed channels. Percentage values in the figure "keys" are flume gradients (from Germanowski and Schumm, 1993).

another 150 hours and sketches of the patterns were made at various times (Figure 3.14). During the 150 hours at the end of the experiment, the flume was uplifted a total of 1.7 cm in three approximately equal increments. The experimental channel was divided into four approximately equal reaches to enable the study of downstream variation in channel characteristics. Two reaches are located above the uplift axis (1, 2) and two reaches are below (3, 4). The channel remained nearly stationary in Reaches 1 and 4, however, significant changes occurred in Reaches 2 and 3, which are immediately upstream and downstream of the axis of uplift.

Sinuosity remained fairly constant for Reaches 1 and 4, indicating that portions of these reaches were not affected by the uplift (Figure 3.15). Reach 2 just upstream of the uplift straightened to a sinuosity of about 1.1 from 450 to 500 hours and then increased back to nearly the original sinuosity by hour 550. The channel cut off several bends above the uplift axis, but these cutoffs were reoccupied by the end of the experiment. Downstream of the uplift, sinuosity increased in Reach 3 during uplift, and it remained high until the end of the experiment.

Meander amplitude, like sinuosity, remained fairly constant in Reaches 1

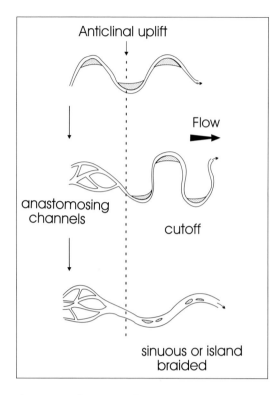

Figure 3.10 Adjustment of a mixed-load meandering channel to anticlinal uplift. Arrows indicate time sequence (from Ouchi, 1985).

and 4 (Figure 3.16). Reach 2 above the uplift axis went through a period of decreasing amplitude with an increase back to the original value after 500 hours. Reach 3 steadily increased amplitude to a maximum at 550 hours.

Meander wavelength remained very constant during uplift for Reaches 1, 2, and 4 (Figure 3.17). Despite changes in both sinuosity (Figure 3.15) and amplitude (Figure 3.16) for Reach 2, wavelength remained constant. Figure 3.14 shows that the meanders above the uplift did not change position. This allowed reduction of meander amplitude in Reach 2 without changing wavelength. However, the wavelength of the steep Reach 3 changed dramatically. The bends in this reach essentially double in size in response to uplift. The channel scoured a completely new course, and it did not reoccupy its old course.

In summary, Reaches 1 and 4 upstream and downstream of the uplift were not greatly affected by the deformation except temporarily. Sinuosity and meander amplitude were significantly affected in Reach 2, but at the end of the experiment Reach 2 had recovered its meandering condition by re-estab-

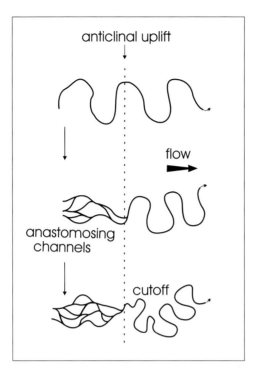

Figure 3.11 Adjustment of a suspended-load meandering channel to anticlinal uplift (from Ouchi, 1985).

lishing its course after a cutoff. Only Reach 3 was dramatically and perhaps permanently affected by deformation. With no further uplift and with a longer time, perhaps even Reach 3 would recover its original pattern.

Confined-river response

If a river is confined laterally by resistant materials, it cannot respond to surficial deformation with lateral changes. The main response will be local aggradation and degradation accompanied by bar changes. Ouchi (1983) experimented with a confined straight channel and the response was very similar to that of the braided channel (Figure 3.2). There was degradation in the central part of the uplift and at the upstream end of the subsidence, and there was aggradation at the upstream side of the uplift and in the central and downstream part of the subsidence.

An experiment was designed by Jin (1983) to determine the effects of locally resistant material on an experimental channel. The purpose was to

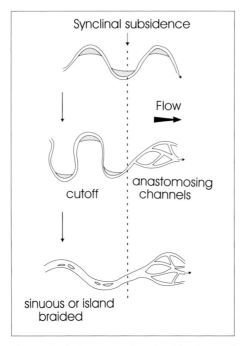

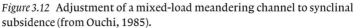

Figure 3.12 Adjustment of a mixed-load meandering channel to synclinal subsidence (from Ouchi, 1985).

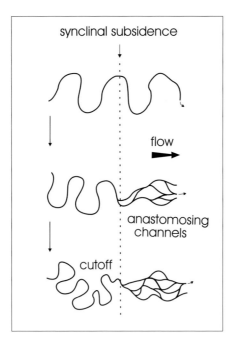

Figure 3.13 Adjustment of a suspended-load meandering channel to synclinal subsidence (from Ouchi, 1985).

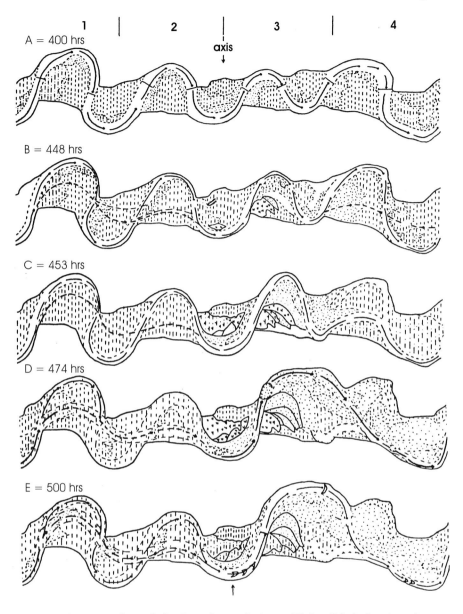

Figure 3.14 Channel planform change during uplift (modified after Jin and Schumm, 1987).

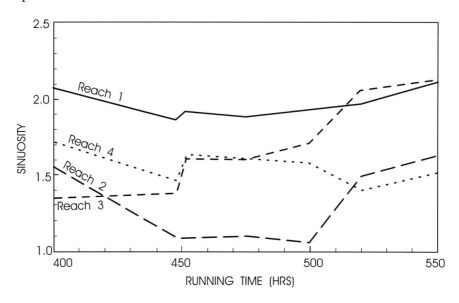

Figure 3.15 Change of sinuosity during uplift for four reaches. Deformation commenced at 400 hours, see Figure 3.14 (modified after Jin and Schuum, 1987).

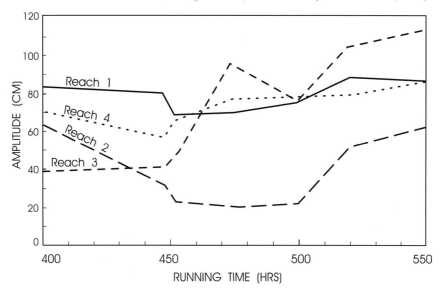

Figure 3.16 Change of meander amplitude during uplift for four reaches. Deformation commenced at 400 hours, see Figure 3.14 (from Jin, 1983).

determine the effect on an alluvial river of the appearance of resistant material in the bed as a result of degradation during uplift. The resistant material was simulated with a block of kaolin clay, that was placed across and under the channel. This block was 10 cm wide, 7 to 8 cm thick, and it was located 4.65 m from the top of the flume at the axis of uplift (Figure 3.18). At this location,

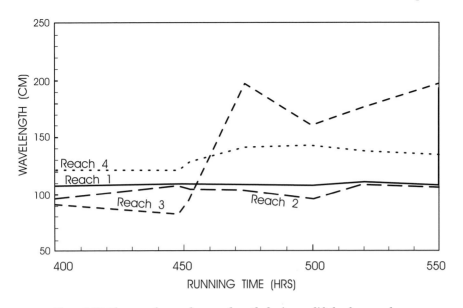

Figure 3.17 Change of meander wavelength during uplift for four reaches.
Deformation commenced at 400 hours, see Figure 3.14 (from Jin, 1983).

the clay block was just upstream of the apex of a bend (Figure 3.19). The stratigraphy of the flume materials was identical to Jin's previous experiment with a 2 cm thick layer of a mixture of 30 percent very fine sand and clay over medium sand with a d_{50} of 0.29 mm. The initial channel had a meandering pattern with a wavelength of 90 cm, an amplitude of 43 cm, and a radius of curvature of 22.5 cm. The experiment was run using a discharge of 0.20 l/sec, which was recirculated in a continuous manner to maintain a constant suspended sediment load. The experiment was run for 740 hours.

The flume was divided into four reaches in order to show downstream changes in channel planform as before. The clay block caused significantly different planform characteristics in the upstream (1 and 2) and downstream reaches (3 and 4). Sinuosity of Reaches 1 and 2 responded in a similar manner to the presence of the clay block, but the response was greater in Reach 2 because it was closer to the block. Sinuosity increased from an initial value of 1.5 to 2.3 for Reach 2 and from 1.5 to 1.95 for Reach 1 (Figure 3.20), because the resistant block prevented meander migration. Sinuosity continued to increase until 400 hours. Reaches 3 and 4 underwent a minor initial adjustment. Reach 4 maintained a very constant sinuosity of about 1.6, while sinuosity of Reach 3 varied slightly about an average of about 1.7. The channel was generally much less active downstream of the clay block than above it.

Meander amplitude increased in a manner very similar to sinuosity (Figure 3.21). The largest increase was in Reach 2 just above the clay block. Amplitude

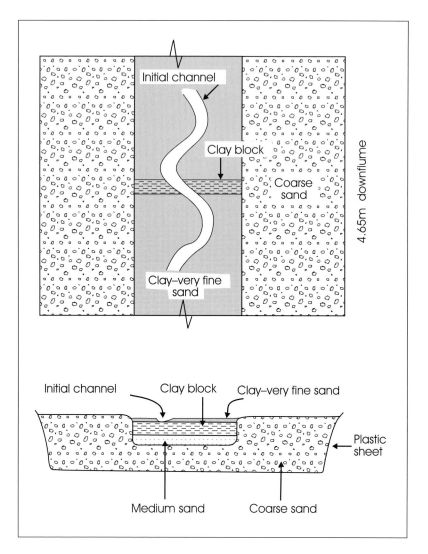

Figure 3.18 Diagram showing the location of the clay block used in the bedrock control experiment (from Jin and Schumm, 1987).

in Reach 1 also increased significantly but not as much as Reach 2. Reaches 3 and 4 underwent minor adjustment and stabilized at an amplitude slightly above the initial value.

Meander wavelength increased for Reaches 1, 3, and 4 and decreased for Reach 2 (Figure 3.22). The decrease in Reach 2 was due to the compressing of meander bends above the clay block, but the increase below the block was due to the meanders shifting downstream away from the block.

Throughout this experiment, the meanders upstream of the clay block

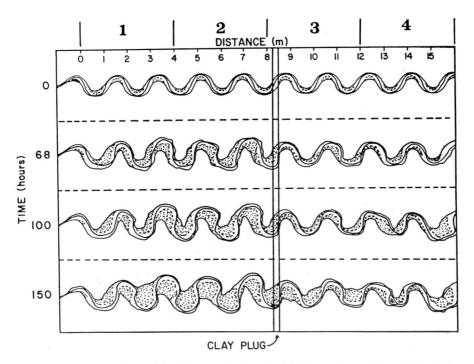

Figure 3.19 Channel development during clay block experiment. Flow is from left to right (modified after Jin and Schumm, 1987).

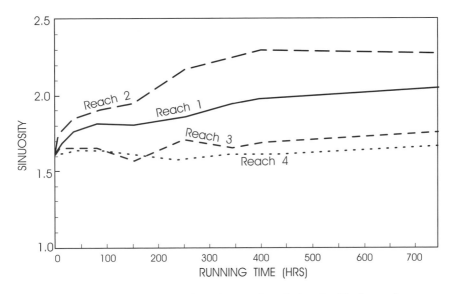

Figure 3.20 Change of sinuosity in four reaches, during clay block experiment (from Jin, 1983).

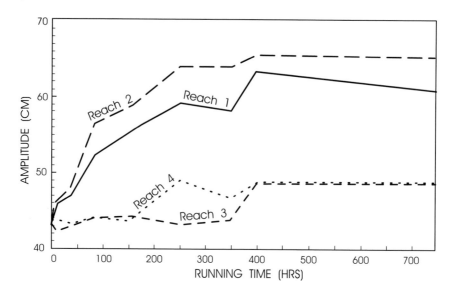

Figure 3.21 Change of meander amplitude in four reaches during clay block experiment (from Jin, 1983).

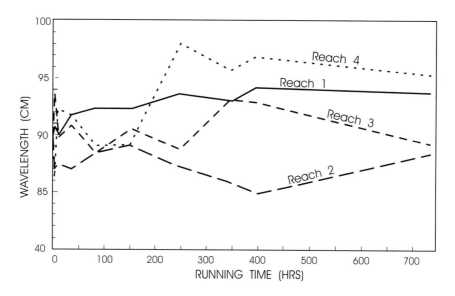

Figure 3.22 Change of meander wavelength in four reaches during clay block experiment (from Jin, 1983).

were being compressed against the block. With continued downstream migration of the bend immediately upstream of the clay block, a neck cutoff of the bend at the block can be expected. With such a cutoff, stream power would increase, perhaps triggering the development of a new bend. Through time, the net effect of the block might be to cause an upstream episodic fluctuation of sinuosity.

Effect of lateral tilting

It seems obvious that if a valley floor is tilted laterally, the channel within that valley will shift laterally or avulse down the valley slope. A series of experiments that involved the effect of doming on a previously formed drainage network provides some information on this matter. The investigation was performed in the Rainfall Erosion Facility (REF) at the Engineering Research Center of Colorado State University (Schumm *et al.*, 1987). The REF is a large rectangular box 9.1 meters wide by 15.2 meters long. It is approximately 1.8 meters deep and has a surface area of 138 square meters. Precipitation was applied through nozzles located approximately 4 meters above the sediment surface.

The REF was filled with sediment that consisted of a range of sand sizes and about 27 percent silt and clay size material. A 5 m diameter inflatable circular plastic bag (snow pillow) was buried at a depth of 0.76 m within this sediment. When water was introduced into the bag, the sediment overlying the bag was deformed. A simple dendritic drainage network was excavated into the sediment surface (Figure 3.23); precipitation was applied, and deformation commenced. Because uplift was domal, the uplifted surface sloped away in all directions, and the channels crossing the dome were displaced laterally as they incised (Figure 3.23). Figure 3.24 shows cross-sections of a channel that shifted to the right away from the center of the dome between 7 and 14 hours. At 14 hours the channel was eroding the right side of the valley, and it left terrace-like features behind as it shifted to the right. These experiments revealed that even during relatively rapid incision, the effect of lateral tilting will be important.

Peakall (1996) investigated the effect of lateral tilt on a stable channel. He used a 5.5 m long 3.65 m wide and 0.5 m deep flume, that contained a range of sediment sizes (mean of 0.21 mm) and which could be tilted up to 3 percent in any direction. The axial gradient was 1.0 percent, and after the introduction of water into a meandering channel, the surface was tilted 0.43, 0.30, and 0.21 percent for three experiments. The result was a shift of the channel down the lateral slope. In addition, sinuosity increased presumably as the amplitude of bends increased down tilt and also because the meanders were not cutoff as frequently as in the no-tilt situation. Peakall cautions that the variability of bend development during any one run makes it difficult to be certain of the conclusions; nevertheless, these changes make good sense physically.

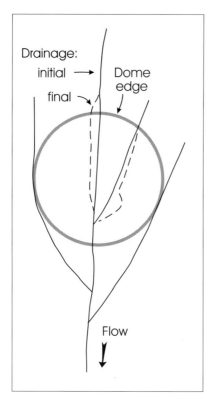

Figure 3.23 Original channel pattern (solid lines) and final position of channels (dashed lines) that shifted away from center of dome (circle) (from Dykstra, 1988).

Effect of deformation rates

Dykstra (1988) observed that the rate at which deformation occurred, significantly affected channel behavior. He compared channel response to rates of doming of 1 cm per 20 minutes and 0.5 cm per 20 minutes. The most rapid rate of uplift caused the least drainage pattern modification. Main channel sinuosity increased, but there was little lateral channel shift because rapid uplift generated major vertical incision. At the slower rate of uplift, significant lateral shift occurred. These results conform to the experimental results of Yoxall (1969), and therefore we conclude that river adjustment to vertical deformation is dependent upon the rate of the uplift, although the total uplift in each case might be the same. Peakall (1996, p. 40) also concluded that there is an effect of rate of tilt on channel response. Rapid rates of tilt can cause avulsion, whereas slow rates are conducive to down-dip migration.

These conclusions appear physically likely, but type of channel (Figure 2.2) may also be very important. For example, the weak banks of bedload channels

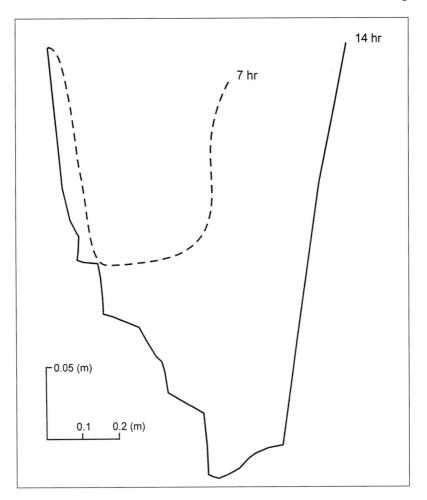

Figure 3.24 Cross-sections showing lateral shift of channel during doming experiment at 7 hours and at 14 hours into the experiment (from Dykstra, 1988). See text for explanation.

may be conducive to lateral shift, whereas the cohesive banks of suspended-load channels may resist shift until avulsion is inevitable.

Effects of unloading

In the previous discussion, the effects of experimental deformation on rivers were considered. Here, we briefly reverse the emphasis and consider the effect of fluvial erosion on deformation, especially on faulting, intrusions and halokinesis (salt tectonics). For example, it has been proposed that progressive denudation would lead to periodic uplift and the formation of

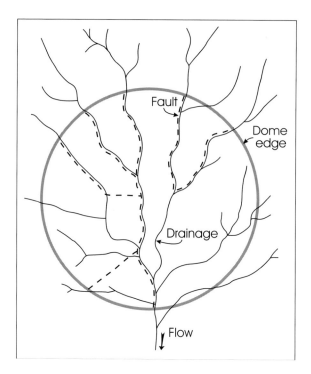

Figure 3.25 Drainage pattern (solid lines) and fractures (dashed lines) that developed during doming experiment. Circle indicates limits of doming (from Dykstra, 1988).

multiple terraces and erosion platforms (Schumm, 1963), as the strength of the earth's crust prevented a continuous isostatic response. For a different opinion, see Gilchrist and Summerfield (1991).

During Dykstra's (1988) experiments, he made observations that indicate the importance of topography on tectonics and unloading on the process of deformation. In the REF, uplift was initiated after a dendritic drainage pattern had formed over the snow pillow. To form this drainage network, precipitation was applied to the REF for 15.5 hours after a 9 cm baselevel drop at the mouth of the REF. The development of the dendritic drainage pattern over the snow pillow indicated that flow was not concentrated due to pre-existing structural features or zones of weakness. The drainage pattern, which developed prior to uplift at 15.5 hours, consisted of two main channels that flowed almost parallel to each other over the snow pillow (Figure 3.25).

At 15.5 hours the snow pillow was inflated, and a dome was uplifted 3 cm. A significant pattern of faulting emerged as a result of uplift. Figure 3.25 shows the fracture pattern and drainage pattern over the dome after this uplift. The upbasin side of the dome exhibited faulting down the center of all of the main valleys. On the downbasin side of the dome, the main fault followed only one main drainage channel.

The pattern of the faulting in the absence of a drainage network was determined during an earlier experiment. The pattern was significantly different from that found in the presence of a drainage network. Uplift in the absence of a drainage network resulted in fractures that radiated outward from a rectangular graben, which formed over the center of the dome. The fracturing exhibited no obvious preferred direction. Therefore, the fracturing shown in Figure 3.25 was influenced by the drainage network, although the channels were incised to a depth of only 7 cm or 11 percent of the total thickness of sediment over the snow pillow. The experiment suggests that relatively insignificant surficial features greatly affect the location of fracturing and release of the tensile stresses that exist in an area of uplift.

The magnitude of uplift over specific areas of a dome was also influenced by unloading. A total of 31 hours of precipitation was applied to the surface of the REF during 15.5 hours with relatively rapid uplift totaling 23 cm. Uplift of the dome was measured at five locations on the dome. Four measurements were made in a square pattern half the distance from the center to the edge of the dome. A fifth measurement was made at the center of the dome. As expected, the center of the dome experienced the greatest uplift. More importantly, by the end of the run, the downstream measurements were 15 percent greater than at the upstream benchmarks, which were at the same relative location on the flank of the dome. During the first 20 hours of the experiment, stream morphology on the upbasin and downbasin sides of the dome was very similar, and uplift rates were similar also during this period. As the run progressed, the streams on the lower half of the dome, however, were beginning to erode more sediment from the dome than was eroded on the upper half. The channel was steeper on the lower half of the dome, which caused significantly greater amounts of sediment to be eroded. Therefore, the lower half of the dome was being unloaded more rapidly than upstream, which caused greater uplift in this area. Therefore, the zone of maximum uplift migrated towards the zone of maximum material removal (Figure 3.26), and the path of the intrusion was changed.

Dykstra's experiments show that the presence of surface irregularities such as valleys can control and lead to further deformation and channel response. For example, river incision into massive bedrock may lead to the formation of large-scale exfoliation joints. These pressure-release joints parallel the canyon walls in the Colorado Plateaus of western United States, and they cause enlargement of the canyons by rockfalls and weathering. Bradley (1963) reports that in massive sandstones, joints that parallel the canyon wall are

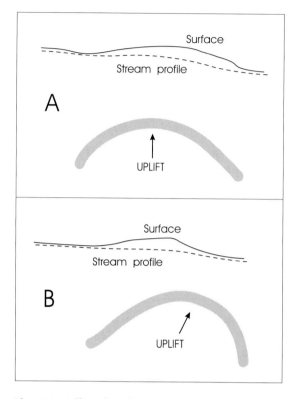

Figure 3.26 Effect of erosion on downstream side of dome on subsequent direction of movement of intrusion. A shows initial conditions and B shows effect of erosion.

most frequent within 7 m of the wall, but beyond 10 m there are no joints parallel to the canyon wall. Obviously, these fractures are due to the excavation of the canyon itself and to the relief of stress in the wall rocks.

At dam sites in the Ohio Valley, Ferguson (1974) reported compressive faults in the valley bottoms and tension fractures in valley walls. The tension fractures are nearly vertical and they parallel the valley walls in a manner similar to those observed in the Colorado Plateaus. The bedrock valley bottoms were subjected to compressive forces produced by fracturing and horizontal expansion of the valley walls. This horizontal force caused failure in the form of arching, thrust faulting, and bedding-plain faults in the valley floor. In the Allegheny Plateaus of eastern United States, the stress release and natural bottom-fracturing and doming were encountered in almost all projects that exposed the sides and bottoms of river valleys. Differential stress, as a result of unloading that is generated by canyon erosion, produced valley anticlines in Canyonlands National Park, Utah and in the central Grand Canyon of Arizona (Harrison, 1927; Potter and Gill,

1978; Huntoon and Elston, 1979). For example, the Meander Anticline follows the course of the Colorado River in the Needles district for many kilometers. Downcutting by the Colorado River initiated the formation of the Meander Anticline. Lateral and upward flow of evaporites into the canyon of the Colorado River caused a sharp upward flexure of the rock layers in the valley walls. In the central Grand Canyon, a system of anticlines lies along the trend of the sinuous course of the Colorado River for a distance of 97 kilometers. These anticlines are an unloading phenomenon, which results from lateral squeezing of saturated shales of the Muav limestone and Bright Angel shale toward the river. The driving mechanism for the deformation is the stress gradient that results from the difference in lithostatic load between the rocks under the 650 meter high canyon walls and the unloaded canyon floor. Saturation appears to weaken the shale sufficiently to allow deformation to take place.

Talbot (1977) and Jackson and Talbot (1986) have demonstrated by experiments that asymmetrical salt domes can be attributed to, among other things, differential loading. They conclude that surface relief variations are a cause of flow in a soft substrate. Hence, erosion is an important control of salt tectonics as well as any intrusion and perhaps faulting as well.

Although this discussion apparently has nothing to do with syntectonic impacts on rivers, it indicates that there can be feedback from fluvial erosion following uplift to the next stage of deformation. That is, erosion induced by uplift can control to some extent subsequent deformation which, in turn, can further impact a river.

Discussion

The experimental studies of river response to syntectonic deformation were performed with different river types (braided, meandering) and with different types and amounts of deformation (uplift, subsidence, lateral tilting, doming). In each case, the alluvial river responded to the deformation in a manner that could be detected in the field. Hence, active tectonics should be recognized by the identification of river anomalies. Nevertheless, when resistant materials are exposed in the bed and banks of an alluvial river, they will significantly affect river response to uplift, and significant changes of sinuosity will result. Also, the rate of deformation appears to significantly influence river response. A rapid rate of uplift will cause incision and prevent lateral shift. However, slow rates of uplift allow more channel adjustment by widening and meandering, and channel positions can change. Hence, resistant materials and different rates of deformation can significantly alter the

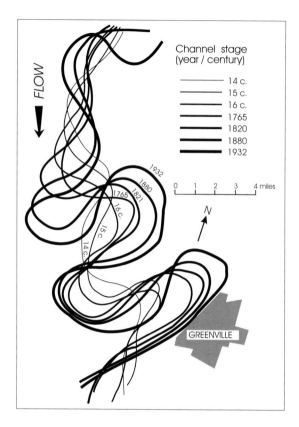

Figure 3.27 Map showing trace of Mississippi River channel centerline near Greenville, MS, years 1765, 1820, 1880, 1932 (from Fisk, 1944).

expected results of deformation, as based on Ouchi's and Jin's experimental results.

Dykstra's (1988) experiments support the experimental results of Jackson and Talbot (1986) and the field observations in the Colorado Plateau and elsewhere (Huntoon, 1982, 1988). Fluvial process can influence fracture patterns and salt and shale tectonics, which in turn further affect channel behavior. In fact, differential erosion of an uplift can cause a change in the direction of the intrusion, which will modify river response.

The results of Jin's experiment with the clay block (Figure 3.19) has significant implications for the interpretation of river response to uplift. The other experiments indicate that a sinuous river will straighten above the axis of uplift or braid as gradient is reduced, and it will become more sinuous downstream of the axis of uplift because gradient is increased. However, upstream of the Monroe Uplift at Greenville, Mississippi (see Figure 4.15 for location) was the most sinuous reach of the lower Mississippi River (Figure

3.27). At Greenville, the river encountered resistant Tertiary-age clay on the upstream flank of the Monroe Uplift. The meanders compressed against this control, as during Jin's experiment (Figure 3.19), to produce high sinuosity upstream of the uplift. Obviously, it is necessary to consider the possibility of such an anomaly in the field.

4

Measured active tectonics

The assumed response of alluvial rivers to deformation, so far, has been based upon logical assumptions (Chapter 2) and experimental results (Chapter 3). Proof, however, must depend upon finding measurements of ongoing deformation and river response to it. In the United States, the National Geodetic Survey (NGS) performs first-order surveys, and these surveys are often repeated, which provides information on surface elevation changes (Figure 1.10). Therefore, there are numerous locations where measured deformation can be related to river response. Several of these locations will be discussed in this chapter to determine if the assumptions of Chapter 2 and the experimental results described in Chapter 3 are meaningful. For example, Gomez and Marron (1991) used NGS surveys to identify a reach of the Belle Fourche River in western South Dakota that is being affected by active tectonics. In this reach, the valley slope, sinuosity, and the floodplain area that was reworked between 1939 and 1981 increased. They conclude that the anomalously large area of floodplain that has been recently reworked in this reach of the Belle Fourche River is a criterion of tectonic activity.

In this chapter, some previously unpublished examples will be presented that demonstrate channel changes that are induced by active tectonics (Figures 4.1, 4.15). In each case, repeat geodetic surveys are proof of active tectonics. The rivers are alluvial except for the Rio Grande and the Guadalupe Rivers, which expose bedrock at the uplift axis. In the discussion of some of these streams and elsewhere in later chapters, the term projected profile will be used to describe a type of longitudinal profile that is a plot of either thalweg elevation or water-surface elevation against valley miles or kilometers. Originally the term projected-channel profile was used (Burnett, 1982; Burnett and Schumm, 1983). It is tempting to refer to these plots as valley profiles, but the valley surface has numerous irregularities due to cutoffs, natural levees, and other depositional and erosional features that are not

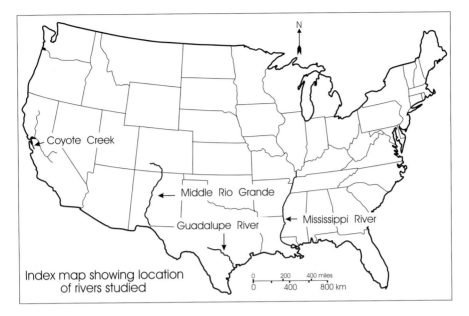

Figure 4.1 Index map showing location of major rivers discussed in this chapter.

related to tectonics and produce a more complicated profile. Therefore, the projected profile is a longitudinal profile of a river with all sinuosity removed.

Coyote Creek

Extremely rapid land subsidence of up to 8.5 m has occurred between 1926 and 1969 in the San Joaquin Valley, Southern California, and rapid subsidence of up to 3.8 m due to ground water withdrawal has also been reported in the lower Santa Clara Valley, California (Tolman and Poland, 1940; Poland and Green, 1962; Poland, 1976). Ground water withdrawal is the major cause of the recent subsidence (Poland *et al.*, 1975). The declining water table reduces pore pressure and increases effective stress, which results in compaction of the sediment.

Coyote Creek is the major stream flowing through the Santa Clara Valley (Figures 4.1, 4.2). The creek passes through what is known as "The Narrows", where it flows over bedrock. Below The Narrows, Coyote Creek flows toward the center of the subsidence, where the slope was steepened the most (Figure 4.2), and it enters the south end of the San Francisco Bay within the area of subsidence. Topographic maps (1/62 500) that were surveyed in 1895 and 1961 were compared to detect channel changes due to the subsidence. Ouchi (1985) divided Coyote Creek downstream from The Narrows into three

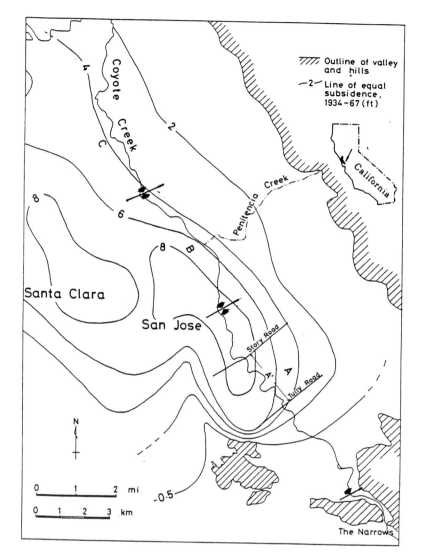

Figure 4.2 Subsidence from 1934 to 1967 in Santa Clara Valley, California. Reaches A, B, and C of Coyote Creek are identified (modified after Poland, 1976).

sections. Sinuosity for each was measured on the topographic maps (Figure 4.3). Sinuosity was greater on the 1961 map in each section, but only in section A and especially in subsection A^1 was there a significant increase of sinuosity, and this occurred where there was a marked steepening of gradient. This conforms to the experimental results, and the pattern changes reflect the channel adjustment to steepening as a result of subsidence.

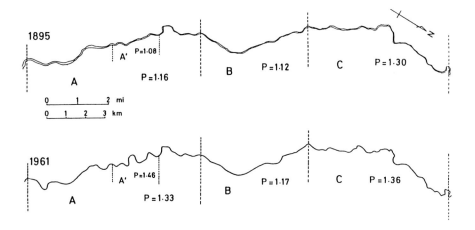

Figure 4.3 Channel patterns of Coyote Creek in 1895 and 1961. Flow is from left to right (from Ouchi, 1983). P is sinuosity.

Rio Grande

In the Rio Grande valley, New Mexico (Figure 4.1), rapid uplift has been measured between Belen and Socorro (Reilinger and Oliver, 1976; Reilinger *et al.*, 1980). The uplift, which was detected by releveling of geodetic survey lines, is a roughly elliptical dome with its center about 25 km north of Socorro (Figure 4.4). The maximum uplift near the center is about 20 cm relative to the periphery, as measured between 1911 and 1951 (5 mm/year). Reilinger and others (1980) suggested that the uplift is caused by expansion of the Socorro magma body (Sanford *et al.*, 1977). The Rio Grande (Figure 4.4) flows across the uplift roughly along its major axis. Late Pliocene alluvium of the ancestral Rio Grande River is uplifted 85 m vertically northeast of Socorro (Bachman and Mehnert, 1978).

Thalweg profiles of the middle Rio Grande River from San Felipe (about 80 km upstream from Belen) to San Marcial (about 48 km downstream from Socorro) all show a large convexity in the uplifted area, between Belen and Socorro (Figure 4.5). Major deposition by the flood of 1929 from Rio Puerco and Rio Salado appears on the 1936–38 profile as a small convexity on the larger convexity from the mouth of Rio Puerco to San Acacia (Figure 4.5). This convexity was removed by 1944, probably by Rio Grande floods in 1937 and 1941. Therefore, local deposition is not the cause of the convexity in the Rio Grande thalweg profile. The fact that the center and areal expanse of the convexity almost perfectly coincides with the uplift indicates that it is caused by the uplift (Ouchi, 1985).

Aggradation, which has been a problem in the middle Rio Grande, is

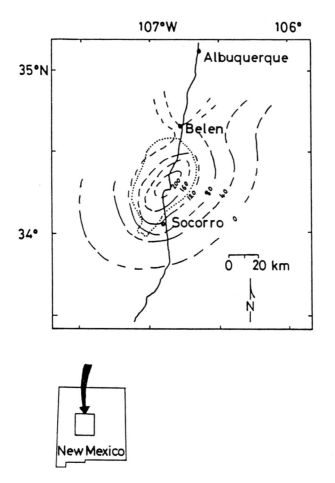

Figure 4.4 Contour map of vertical crustal movement in the Albuquerque – Socorro area, New Mexico (modified after Reilinger *et al.*, 1980). Contours represent deformation between 1934 and 1958 in mm.

caused by the uplift (Figure 4.5). There is aggradation from at least 1918, except from the mouth of Rio Puerco to Lemitar where, in fact, degradation can be expected, as the river adjusts to uplift. Aggradation can easily be explained by the slope reduction of the upstream side of the uplift (Figure 4.5). The reduced discharge of the Rio Grande and increased sediment supply from tributaries since the late nineteenth century accelerated this aggradation. In the reach from Lemitar to the downstream limit of the uplift, sediment derived from the uplift has caused aggradation (Figure 4.5).

The Rio Grande has responded to uplift. There are terraces in the central area of uplift that display marked upward convexity. There is aggradation upstream and downstream of the convexity and degradation in the central to

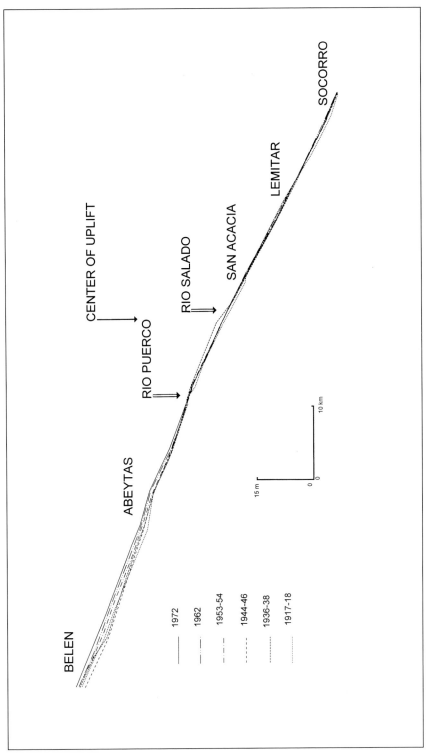

Figure 4.5 Longitudinal profiles of the Rio Grande from Belen to Socorro (from Ouchi, 1983).

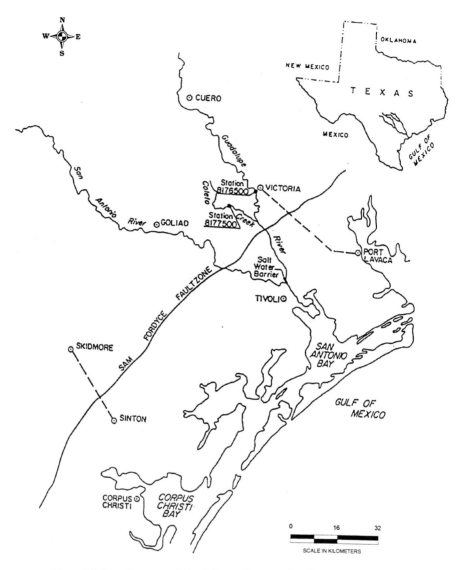

Figure 4.6 Location map of Guadalupe River and leveling lines between Victoria and Port Lavaca, and between Skidmore and Sinton, Texas.

downstream part of the convexity. These observations agree well with the results of Ouchi's experimental study of the effect of uplift on braided streams (Figures 3.2, 3.3).

Guadalupe River

The Guadalupe River rises in south-central Texas (Figure 4.1) and flows to the southeast through Cuero and Victoria, to its mouth at the head of

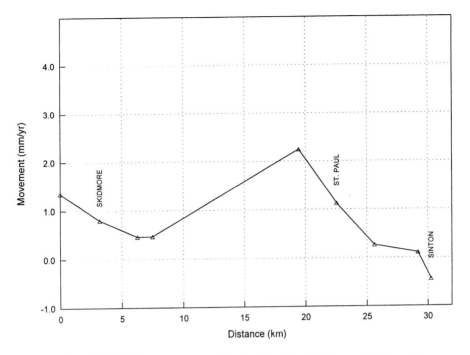

Figure 4.7 Relative movement on leveling line from Skidmore to Sinton, Texas.

San Antonio Bay (Figure 4.6). In the vicinity of Victoria, Texas, the river crosses the Sam Fordyce Fault zone, which extends along the Texas Gulf Coast. Two NGS leveling lines cross the structure (Figures 4.7, 4.8), and both lines indicate relative subsidence immediately below the fault zone.

A detailed field study of the river was undertaken to collect morphologic and sedimentologic data in order to evaluate river response to active tectonics (Schumm *et al.*, 1984). Twenty-one cross-sections were surveyed along the Guadalupe River between the mouth and Cuero, Texas (Figure 4.6). The channel and average bank profiles for the lower 65 valley kilometers of the Guadalupe River show direct evidence for active movement along the Sam Fordyce Fault zone (Figure 4.9). Valley kilometers are measured along the valley in contrast to measuring channel lengths. A sharp break in both thalweg and bank profiles between kilometers 29 and 31 occurs at the fault zone. The profiles immediately upstream of the fault zone are steep relative to the reach downstream, but no bedrock control was observed in the field. The slope breaks are interpreted to be the result of the fault zone. The downstream fault block is moving in a southeast direction toward the coast. The reach immediately downstream of the fault zone is backtilted in order to fill the void created by movement along this listric fault.

Channel cross-section morphology is very different on opposite sides of the

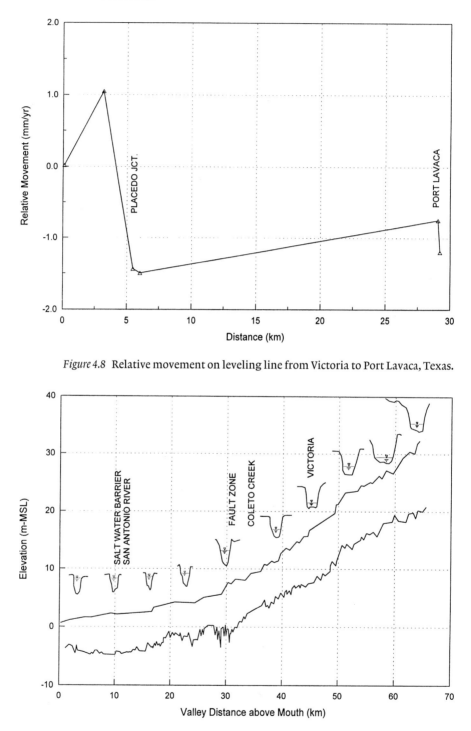

Figure 4.8 Relative movement on leveling line from Victoria to Port Lavaca, Texas.

Figure 4.9 Thalweg and average bank profiles for Guadalupe River from a 1961 survey. Also shown are cross-sections from the 1988 survey.

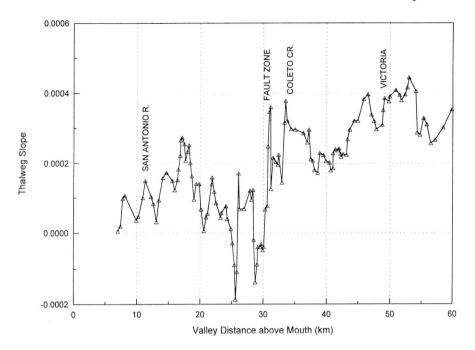

Figure 4.10 Ten-point moving average of thalweg slope for the Guadalupe River, calculated from the 1961 thalweg profile data.

slope change at km 30. Downstream, the channel is relatively small with a uniform width, and flows observed during field investigations were nearly bankfull (Figure 4.9). Upstream, the cross-sections are much wider and deeper, and observed flow was well below bankfull (Figure 4.9). The channel cross-sections also vary more in size and shape upstream of the fault zone.

The thalweg slope also reflects movement along the fault zone (Figure 4.10). There is a general increase in slope in the upstream direction, typical of a concave channel profile; however, this trend is disrupted downstream of the fault zone where the slope is actually negative (Figure 4.10). The irregular thalweg slope profile indicates that the Guadalupe River is adjusting to active movement along the fault.

The sinuosity of the Guadalupe River shows variations that are also spatially related to the fault zone. An abrupt change in channel sinuosity occurs near the fault zone (Figure 4.11) (Ouchi, 1985). Sinuosity increases from about 1.7 to about 2.5 from km 27 to 32 in response to the steepened slope upstream of the fault zone. A very low channel gradient downstream is responsible for the relatively low sinuosity. The low sinuosity above km 50 is due to bedrock control.

Downstream changes in channel width, depth, and cross-section area of the

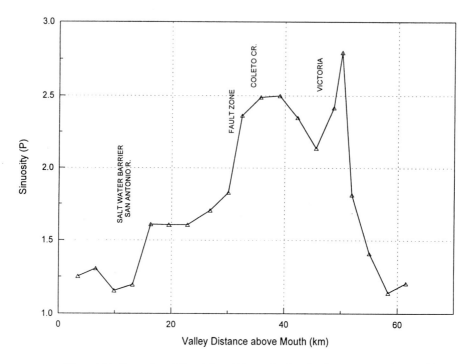

Figure 4.11 Sinuosity of Guadalupe River calculated at increments of two valley miles.

Guadalupe River are associated with the Sam Fordyce Fault zone. These variables generally decrease in the upstream direction, but there are variations in the relations that can be attributed to active faulting. For example, the channel top width, bottom width, and water surface width all locally increase at the fault zone (Figure 4.12, km 29). There is also an increase in channel depth in the immediate vicinity of the fault zone (Figure 4.13) and therefore, cross-section area also increases at the fault (Figure 4.14). The variations in each of the variables constitute a strong argument in favor of active movement along this fault, as indicated by the NGS repeat surveys (Figures 4.7, 4.8). The coincidence of channel changes with the fault location indicates that active movement of the downstream (hanging) block of the Sam Fordyce Fault is causing local discontinuities in the morphology of the Guadalupe River.

Rivers of the Gulf Coast

Several rivers in Louisiana, Mississippi, and Alabama provide an example of river response to the Monroe and the Wiggins Uplifts (Figure 4.15). Field investigations were conducted along several rivers located on the Gulf Coast extending from Alabama to Texas (Schumm *et al.*, 1984). The rivers

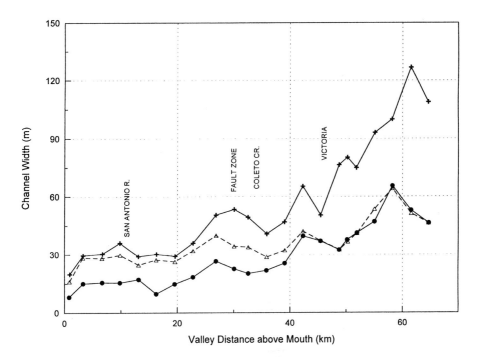

Figure 4.12 Channel top width (+), bottom width (●), and water surface width (△) of Guadalupe River from cross-sections surveyed in 1988.

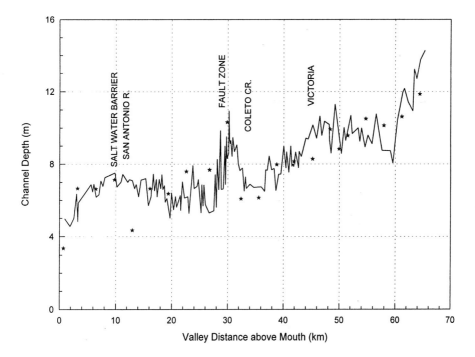

Figure 4.13 Channel depths of Guadalupe River, calculated from the 1961 survey (solid line) and from the 1988 survey data (stars).

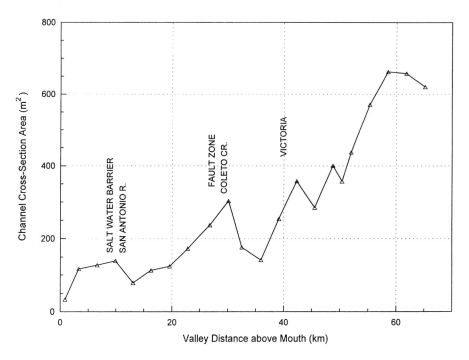

Figure 4.14 Channel cross-sectional area of Guadalupe River from 1988 survey.

include, from east to west, Tallahala Creek and Bogue Homo, which cross the Wiggins Uplift in Mississippi, Mississippi River (Figure 4.1), and five streams crossing the Monroe Uplift in Louisiana.

The Gulf Coast of southeastern United States extends from Florida westward to the Mexican border. It has little topographic variability, and geologic structures are subtle. Submergence is generally assumed to be prevalent, but Clark *et al.* (1978) suggest slight post-Pleistocene emergence of the coast (Figure 1.11), and Jurkowski *et al.* (1984) identify up-arching of portions of the Gulf coastal plain (Figure 1.10). Otvos (1981) has recognized a series of faults that parallel the coast and cause deflection of stream courses.

The Gulf Coast (Murray, 1961) is composed of relatively undeformed wedges of Mesozoic and Cenozoic sediments, which dip and thicken towards the coast indicating subsidence over geologic time. Quaternary deposits are fluvial, deltaic, and marine sediments of varying lithology. Structures within the Gulf Coast include the Ocala Uplift in northeastern Florida, the Apalachicola Embayment extending south to the coast from the southwest corner of Georgia, and the Wiggins Uplift (Figure 4.15) in southern Louisiana. Between Louisiana and Florida, coastal terraces have been identified and related to changes in sea level. Recent studies (Otvos, 1981) indicate a tectonic

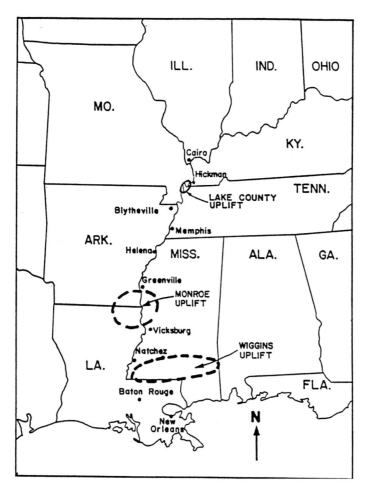

Figure 4.15 Index map of lower Mississippi River valley.

origin for several of these terraces. Offshore subsidence, and loading of the continental shelf by the Holocene sea level rise, could cause an onshore bulge analogous to the forebulge flanking the Pleistocene ice sheets (Clark *et al.*, 1978, p. 272). Nunn (1985) notes that the Baton Rouge fault has displaced pavements and foundations in Baton Rouge, Louisiana and that the coast is in a state of tensional stress as a result of sediment loading.

Wiggins Uplift

Holdahl and Morrison (1974) summarize the evidence from precise NGS resurveys (Figure 4.16), which place a zone of moderately active uplift in southern Mississippi and southwestern Alabama. Relative movement along five leveling lines in southeastern Mississippi and southwestern Alabama indicate that the Wiggins Uplift is active and the rate of uplift is 4 mm/yr (Figures 4.16, 4.17).

Figure 4.18 shows releveling profiles along a survey line from Jackson, Mississippi, to New Orleans, Louisiana that crosses directly over the axis of the Wiggins Uplift. The apparent axis of uplift is located south of Laural, Mississippi (Figure 4.17). The maximum uplift rate is 3.3 mm/yr. The anomalously large value at km 135 has been discounted, as it is probably caused by localized benchmark movement not associated with the broad-scale upwarping. However, the historic rate of movement is much greater than that suggested by Tertiary deformation, which implies that the uplift may be episodic. Several other survey lines are located near the Wiggins Uplift, and all indicate uplift.

Three alluvial streams, which flow southward over the Wiggins Uplift in central Mississippi, were chosen for analysis to determine channel response to these recent movements. The streams are from west to east, the Pearl River, Tallahala Creek, and Bogue Homo (Figure 4.17). All three streams carry substantial amounts of sand and gravel.

Several terraces were identified on topographic maps (Figures 4.19, 4.20, 4.21). The heights of these terraces range from 1.5 to 49 m above the present channels and they are Pleistocene to Holocene in age. The highest terraces are found along the Pearl River valley, ranging in height from 6.6 to 49 m above the present channel. Lower terraces ranging in height from 1.5 to 17 m above the channels are found in all valleys, and all valleys contain an active floodplain. This surface is defined as the lowest surface in each of the valleys, which ranges in height from 1.2 to 5 m above the low-water channel.

Broad upward convexities are obvious on all of the terrace profiles (Figures 4.19, 4.20, 4.21). For each of the profile sets, the crests, or axis, of the convexities coincide. This indicates that the axis of deformation has remained in the same location. For example, the Pearl River valley has an apparent uplift axis at about km 40 from the mouth, and the extent of the upward convexity is from km 106 to km 176.

Note that in each of the figures showing longitudinal profiles, the lowest is identified as a projected profile as defined earlier (Figures 4.19, 4.20, 4.21). As described earlier, these projected profiles are not affected by changes in sinuosity because the thalweg elevation at a given location is plotted against valley distance rather than channel distance (Burnett, 1982).

For each of the valleys, a comparison between recent vertical movement and the terrace profiles shows that the profile convexities correspond with recent uplift rates (Figures 4.19B, 4.20B, 4.21B). Also, the location of the axis

Units for Contour Levels are mm/yr.

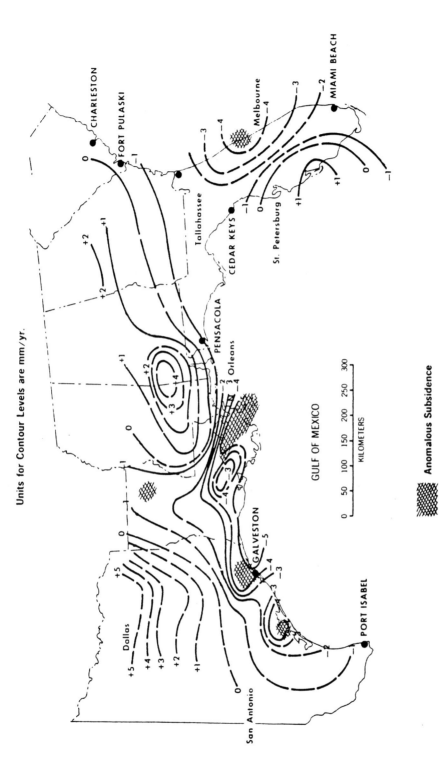

GULF OF MEXICO

KILOMETERS

0   50   100   150   200   250   300

Anomalous Subsidence

Figure 4.16 Rates of elevation change of Gulf Coast in mm/year (from Holdahl and Morrison, 1974).

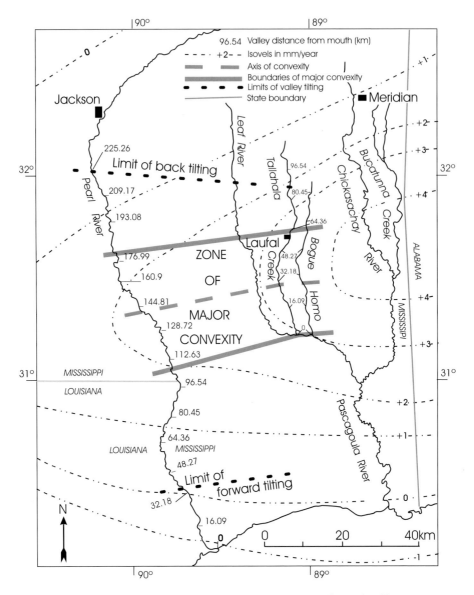

Figure 4.17 Boundaries and axis of surface upwarping, determined by terrace deformation, compared to isovels of historic vertical benchmark movement (from Burnett, 1982).

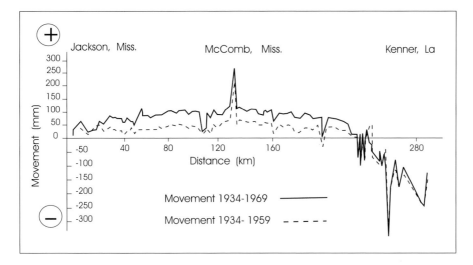

Figure 4.18 Vertical benchmark movement along Jackson, Miss. – New Orleans, La survey route (from Burnett, 1982).

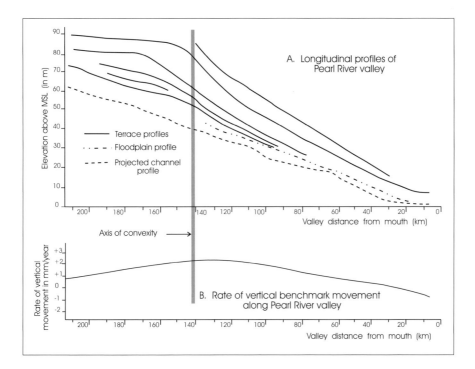

Figure 4.19 Longitudinal profiles of Pearl River valley compared to rate of vertical benchmark movement (from Burnett, 1982).

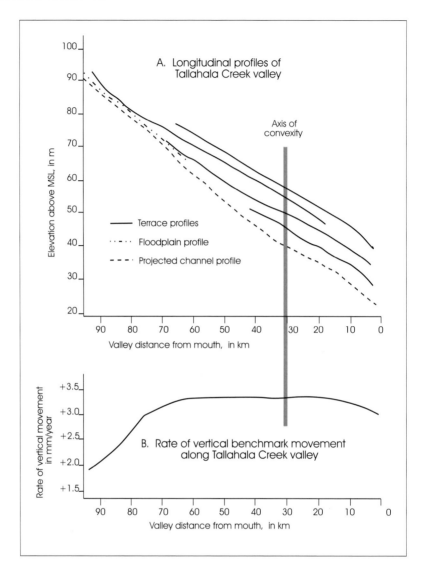

Figure 4.20 Longitudinal profiles of Tallahala Creek valley compared to rate of vertical benchmark movement (from Burnett, 1982).

of upwarping, in each case, is associated with the maximum rate of recent uplift. The boundaries and axis of upwarping, as interpreted from the profile convexities, are shown in relation to the isovel map of recent vertical movements (Figure 4.17). These observations are evidence that the profile convexities are caused by active tectonics.

Pearl River

The Pearl River is the largest and the westernmost of the streams studied (Figure 4.17). It has a valley length of over 350 km and an average

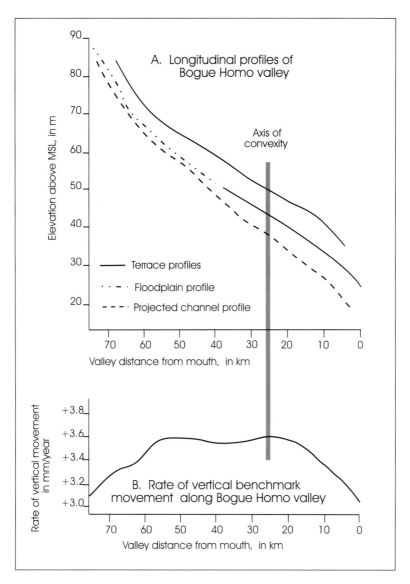

Figure 4.21 Longitudinal profile of Bogue Homo as compared to rate of vertical benchmark movement (from Burnett, 1982).

annual discharge of 204 m³/s near Columbia, Mississippi. Benchmark movement (Figure 4.19B) shows that forward tilting of the valley is occurring between km 32 and 140; and back-tilting is occurring from km 141 to 216. Major valley upwarping apparently has occurred between km 106 and 176. In the backtilted portion of the valley, the channel is entrenched below the lowest terrace surface. This surface has ridge and swale topography, which indicates that it formed as a floodplain by lateral migration of the channel.

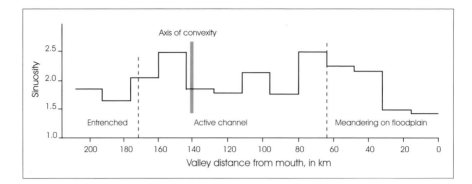

Figure 4.22 Ten-mile average of Pearl River sinuosity (from Burnett, 1982).

However, it is no longer active due to entrenchment of the channel, and a lower floodplain has not yet developed. Burnett (1982) reports outcrops of Tertiary clay along the base of the banks in several locations indicating entrenchment to bedrock.

The Pearl River channel can be divided into three major reaches, which can be related to channel response to the recent valley upwarping. The lower zone (km 16–64) is characterized by a relatively stable meandering channel and a well-developed floodplain. This reach is in the lower portions of the forward tilted zone, and the channel has adjusted to the steepened gradient by an increase in sinuosity (Figure 4.22). The middle reach (km 64–171) is characterized by a very active and unstable channel with numerous recent meander cutoffs, active point bars, and mid-channel bars. This reach is also marked by active erosion, which involves renewed channel scour and remobilization of bed and bank materials. An active floodplain is presently being developed in this reach 3 to 4.5 m above the channel. The upper reach (km 171–220) is characterized by a recently entrenched and confined channel with low sinuosity. An active floodplain has not yet developed, and the channel does not show either lateral migration or extensive bar development. Apparently backtilting and recent entrenchment have caused reductions in channel slope and stream power, thus producing a low sinuosity pattern (Figure 4.22) associated with a low valley slope (Figure 4.19).

Channel incision has occurred throughout the entire upwarped area of the Pearl River valley because stream power is large, and the river responds very quickly to periods of upwarping. Therefore, the projected channel profile does not show a marked upward convexity associated with the upwarp (Figure 4.19) because the river has been able to downcut and maintain its course through the uplift.

Tallahala Creek

Tallahala Creek is located approximately 80 km to the east of Pearl River, and it flows generally north to south across the uplift (Figure 4.17). It is significantly smaller than the Pearl River with a valley length of 112 km and an average annual discharge of 27 m³/s.

As shown by deformation of the terrace profiles and by releveling survey data (Figure 4.20), forward tilting of the Tallahala Creek valley has occurred between about km 0 and 30. Backtilting has occurred between km 50 and 86. Tallahala Creek has significant variations in channel characteristics, which appear to be directly related to channel response to uplift. An upper reach above km 68 is characterized by a very broad low-level floodplain containing a meandering low-water channel and a network of flood channels, which form an anastomosing pattern during high-water periods. The floodplain is up to 5 km wide, and it is only 1.5 to 3 m above the low water channel.

The channel slope (Figure 4.23A) in this upper reach ranges from 0.5 to 1.1 m per km. The channel flows entirely over alluvium composed of sand and gravel, and no Tertiary bedrock is visible. The channel has not entrenched, and it is not actively degrading its bed. This channel is only 9 m wide with a gentle gradient, and it probably floods even during small runoff events. Apparently, this upper reach is dramatically affected by valley back-tilting. The reduced gradient has caused a reduction in channel capacity and stream power, and an anastomosing channel pattern has developed.

A middle reach between km 50 and 68 is characterized by a single narrow and deep channel that is confined by steep 8 m high banks. The terrace surface above the banks is continuous with the floodplain surface upstream (Figure 4.20). Therefore, this surface is now perched, and it is no longer active. In addition, a lower floodplain has not yet developed. The channel has entrenched about 6 m below the former floodplain. In this middle reach (km 50 to 68), there is an increase in the valley slope (Figure 4.23A). The channel in this reach has a relatively low sinuosity (Figure 4.23B). Channel straightening is occurring, and the insides of bends are being severely eroded while the outsides appear relatively more stable. A high frequency of recent cutoffs is also evident.

In the lower reach downstream of km 50, the channel is characterized by active point bar and mid-channel bar development and a prevalence of chute cutoffs and the channel is entrenched well below the lower terrace level (Figure 4.20). The average sinuosity is very low (1.4–1.6) compared to upstream (Figure 4.23B).

In conclusion, backtilting above the axis has caused the development of an anastomosing pattern. Forward tilting in the lower 30 miles of the valley also

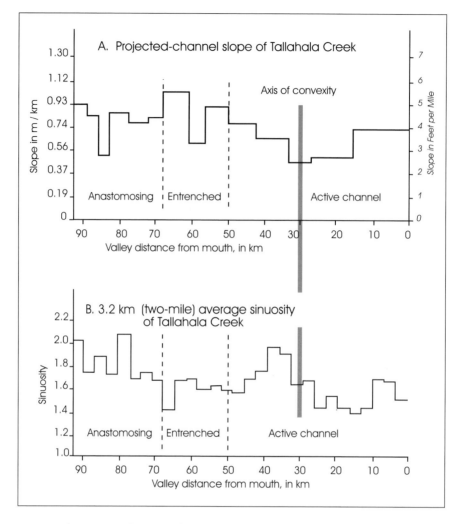

Figure 4.23 Projected profile slope and average sinuosity of Tallahala Creek (from Burnett, 1982).

has initiated channel scour, nickpoint development, and consequently channel entrenchment, which has progressed upvalley through the axis to about km 68. Upstream of this point the channel has not yet entrenched. Note especially that the upstream floodplain is a terrace downstream (Figure 4.20), and between them the floodplain is absent.

Bogue Homo

The Bogue Homo is the smallest of the three streams studied, with an annual average discharge of 18 m³/s and a valley length of 88 km. The valley is 11 km to the west of and generally parallels the Tallahala Creek valley (Figure

4.17). As shown by deformation of the terrace profiles and by releveling survey data (Figure 4.21), forward tilting of the valley has occurred from the mouth to km 26, and backtilting has occurred between km 26 and 60.

From km 45 to the headwaters, the channel is generally confined by a narrow valley cut into Tertiary sediments, and it has a typical concave-up profile. This portion of the channel is not alluvial, and it has a relatively steep valley gradient so it is not readily affected by the valley tilting. Along the lower 45 km of the valley, however, Bogue Homo is predominantly an alluvial channel. By examining the various patterns of channel characteristics in this reach, the lower 45 km may be divided into three basic reaches that are very similar in nature to the upper, middle, and lower reaches of the Tallahala Creek.

The upper reach, between km 32 and 46, has a broad floodplain 1.5 to 3 km wide and 1.5 to 2.5 m above the low water channel. Numerous interconnecting flood channel depressions are evident across the entire floodplain, which are similar in appearance to those in the upper anastomosing reach of Tallahala Creek. The low-water channel also shows similar characteristics and a high average sinuosity (1.7 to 1.8) (Figure 4.24). This high sinuosity is probably caused by the irregular nature of the channel pattern, as it flows through the network of floodplain channels. Similarly, the development of point bars, mid-channel bars, and meander cutoffs is minimal. However, numerous side channels and tree-covered islands have developed in association with the anastomosing pattern. This upper reach is immediately above the uplift axis on the backtilted portion of the valley. Apparently reduction in the valley gradient has facilitated the development of these low-energy channel characteristics.

Between km 20 and km 32, a single confined channel has developed due to entrenchment from 4.5 m to 5.5 m below the floodplain. This surface now forms a lower terrace along this reach (Figure 4.21), and a lower active floodplain has not yet developed. Entrenchment of the meanders into the underlying Tertiary units has caused high channel sinuosity along this reach (Figure 4.24). However, channel straightening is evident by the high frequency of recent cutoffs. In addition numerous mid-channel bars have developed causing local braided reaches. This reach is analogous to the middle reach of Tallahala Creek.

The lower 20 km of the Bogue Homo valley have similar channel characteristics to those in the lower reach of Tallahala Creek. The channel shows active lateral migration accompanied by valley widening, recent floodplain development, and active point bar growth. Also, numerous mid-channel bars and chute cutoffs have developed to form local braided reaches. Similar to the

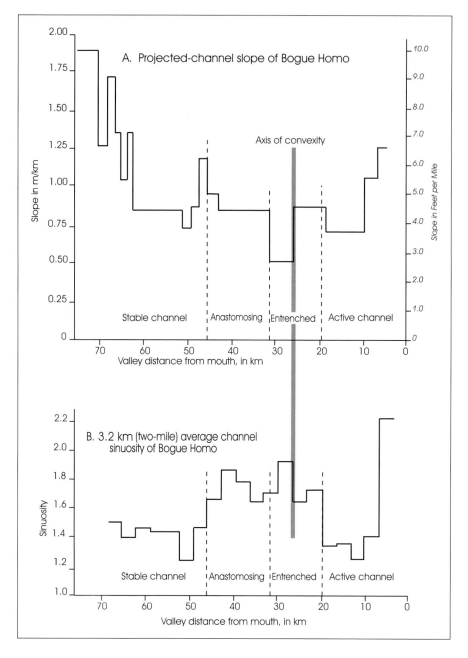

Figure 4.24 Projected profile slope and average sinuosity of Bogue Homo (from Burnett, 1982).

Tallahala, the Bogue Homo appears to be readjusting to recent channel entrenchment and to forward tilting of the valley. Note that again, the upstream floodplain becomes a downstream terrace (Figure 4.21).

In conclusion, the Bogue Homo has responded to the uplift in the same manner as the Tallahala Creek. However, Bogue Homo is smaller, and it should respond more slowly to the valley tilting. This appears to be the case since channel entrenchment has not progressed much beyond the uplift axis, and the lower reaches of Bogue Homo are in a less advanced stage of readjustment to the forward tilting and entrenchment. In the lower reaches the channel is generally more confined, and an active floodplain is less well developed.

Bedrock occurs in Bogue Homo and Tallahala Creek, at the uplift axis, which makes it difficult to distinguish between the effects of bedrock control and active deformation. However, the bedrock exposure is a result of active uplift and channel response.

Monroe Uplift

The Monroe Uplift, the extent of which is defined by deformed Cretaceous and Tertiary strata, is a dome approximately 120 km in diameter. It is situated mostly in northeastern Louisiana (Figures 4.1, 4.25), but it extends into southeastern Arkansas and west central Mississippi (Wang, 1952). Its eastern-most extension includes the Mississippi River between Greenville and Vicksburg. Geological evidence that the Monroe Uplift was active since the Tertiary has been presented by several authors. Veatch (1906) discussed the existence of two active linear structures, which pass through the uplift. He further claimed that recent movement along the west end of the flexure has resulted in the formation of a series of shoals on the Sabine and Angelina Rivers and swamping of an area in the Angelina River valley in eastern Texas (Veatch, 1906).

The uplift consists of the Boeuf and Tensas basins as well as Macon Ridge and the Mississippi River. The basins are composed of large backswamp areas crossed by several old Arkansas River channels and well-developed natural levees. The Macon Ridge in the center of the Uplift is composed of five mid-Wisconsin glacial outwash terraces (Saucier and Fleetwood, 1970), which form a narrow, north–south trending, elongated area of high ground. Several streams – Ouachita River, Bayou Bartholomew, Boeuf River, Big Colewa Creek, Bayou Macon, and Deer Creek (Figure 4.25) – that cross the Monroe Uplift in northeastern Louisiana and which generally parallel the Mississippi River were studied by Burnett (1982). They flow generally to the southwest across the uplift and they locally occupy old abandoned courses of the Arkansas River.

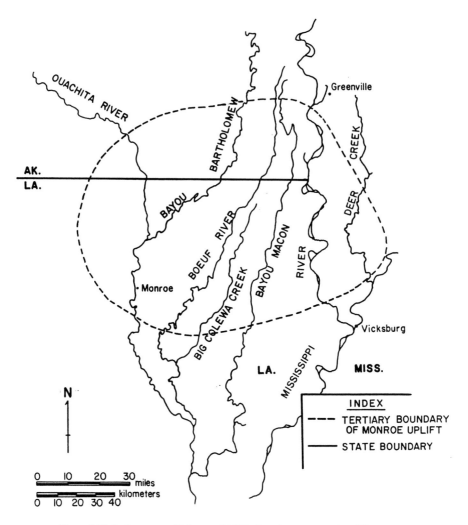

Figure 4.25 Index map of Monroe Uplift showing extent of uplift (modified after Burnett and Schumm, 1983).

The most recent surface movements are indicated by precise resurveys. The geodetic surveys do not cross the uplift axis; however, they indicate uplift of the southern part of the area between 1934 and 1966. The longitudinal profiles of Pleistocene and Holocene terraces show convexities, which are due to uplift. The Monroe Uplift is still active today and modern stream and valley-floor profiles exhibit the effects of the uplift similar to those shown by terrace profiles. Indeed, valley-floor profiles of the Monroe Uplift streams show an obvious zone of upward convexity (Figure 4.26).

Sinuosity for the five rivers is plotted with reference to the position of the uplift axis (Figure 4.27). In each case sinuosity increases as the axis is

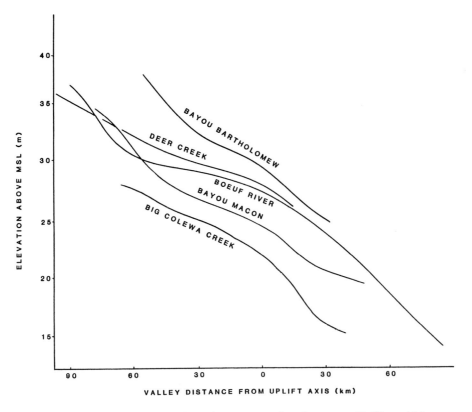

Figure 4.26 Projected profiles of streams crossing the Monroe Uplift, and Macon Ridge, which is a remnant of a Pleistocene Mississippi River terrace (modified after Burnett and Schumm, 1983).

approached or crossed. Where sinuosity is high above the area of uplift, there is a decrease of sinuosity as the axis is approached and an increase downstream of the axis of uplift (Bayou Bartholomew). This behavior conforms to experimental results.

The Big Colewa Creek channel can be used to summarize the effect of uplift on these alluvial channels. It can be divided into three reaches along its valley (Figure 4.28A). The lower reach, from the mouth to km 24, has a high average bank height of about 6 m. The middle reach, from km 24 to 56, shows a clear upstream decrease in the average bank height from 6 to 2 m. The upper reach, above km 56, has a constant low average bank height of 2 m. Degradation may have occurred in the lower reach, whereas entrenchment is still in progress in the middle reach, and in the upper reach entrenchment has not yet occurred. Figure 4.28B shows changes in valley slope and channel thalweg slope with valley distance. The valley slope remains high from the mouth to km 32 and then it suddenly decreases. The break in slope, at km 32, defines the apparent

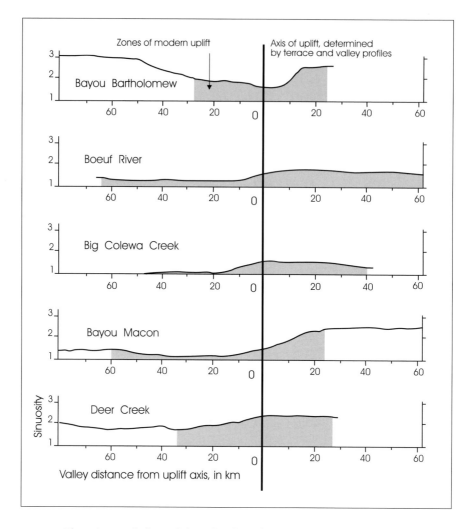

Figure 4.27 Variations of sinuosity along five Monroe Uplift streams. Distance above and below uplift axis is shown. Downstream is to the right (from Burnett, 1982).

location of the uplift axis. The thalweg slope is high from the mouth to km 48 and, then, it too suddenly decreases. The fact that the two curves do not coincide suggests that the channel has incised through the axis of uplift.

Sinuosity is approximately 1.2 in the upstream reach of Big Colewa Creek (Figure 4.28C). Downstream between km 45 and 40, sinuosity increases to about 1.7, and then it gradually decreases to 1.5 at the mouth.

In summary, numerous relations between the morphology of the streams and terraces and the underlying Cretaceous and Tertiary structures of the Monroe Uplift indicate and NGS data demonstrate that the area is still active

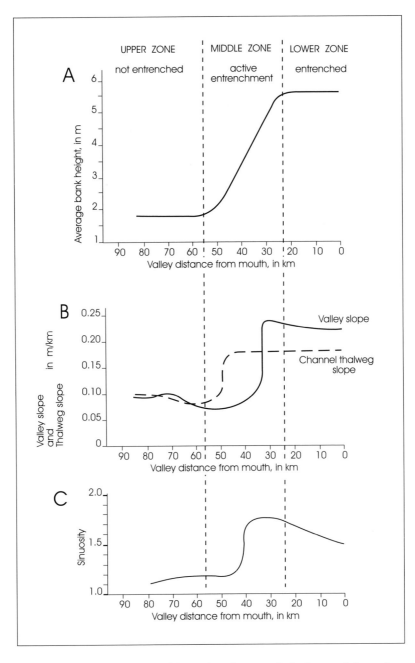

Figure 4.28 Effect of Monroe Uplift on Big Colewa Creek. A, change of channel depth. Three reaches show different degrees of response to uplift. B, change of channel gradient and slope of valley floor. C, change of sinuosity (modified after Burnett and Schumm, 1983).

tectonically, and the patterns of channel changes and present stream morphology support the experimental results of Schumm and Khan (1972), Ouchi (1985), and Jin (Jin and Schumm, 1987).

Discussion

All of the rivers discussed in this chapter are affected by measured deformation. Confidence can be placed in the evidence for uplift on the Wiggins and Monroe Uplifts and in the Rio Grande Valley and subsidence affecting Coyote Creek. In each case, deformation is relatively rapid and measurable and the rivers respond as expected when compared to theory and experimentation (Chapters 2 and 3). These results permit a significant degree of confidence to exist when river anomalies are detected and explained as a result of tectonic deformation.

Although emphasis has been placed upon variations of planform, it is apparent that other channel characteristics are altered by active tectonics. Gomez and Marron (1991) describe changes in floodplain and channel gradient, sinuosity, floodplain reworking and channel depth, as a result of deformation and the examples of the Gulf Coast streams support their findings.

The streams of the Gulf Coast indicate that, as expected, larger streams more readily adjust to deformation. Therefore, perhaps attention should be directed toward the smaller channels as indicators of active tectonics. For example, Bogue Homo is smaller than Tallahala Creek, and the size difference has evidently resulted in varying degrees of entrenchment of each creek. Tallahala Creek has undergone significantly more erosion than Bogue Homo, as illustrated by the depth of the channel through the uplift axis and the quantity of exposed bedrock along the valley course. This evidence implies that Tallahala Creek has adjusted to active uplift to a greater extent than Bogue Homo, presumably due to greater discharge and higher stream power. Variations in local uplift rate may also be important. Although Tallahala Creek may have responded to uplift more than Bogue Homo, both streams presently appear to be in a phase of active adjustment. However, the larger Pearl River appears to be farther along with adjustment to uplift (Figure 4.29). It has incised through the axis of uplift and has a well-developed floodplain downstream. Therefore, the condition of a channel on an uplift will depend on size as well as other factors.

Even along the largest rivers, uplift will affect flood hydrology. For example, analysis of high-water marks along Tallahala Creek showed that flooding is more likely to occur upstream of an uplift axis, but flooding will be less frequent at and downstream of the axis. During the great flood of 1927 in

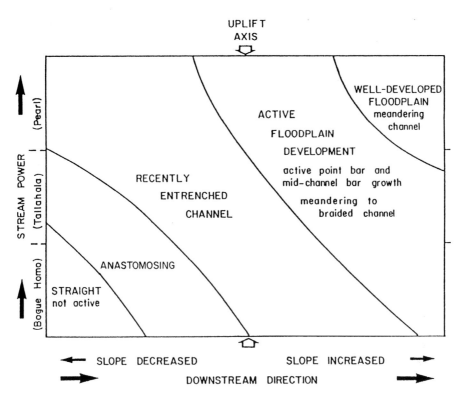

Figure 4.29 Alluvial-river channel characteristics in relation to relative stream power and location relative to the axis of active uplift. Note that as stream power increases, the zone of active channel entrenchment moves upstream through the axis of uplift indicating a more complete response to the uplift by larger streams (from Burnett, 1982).

the Mississippi Valleys 5 100 000 hectares of land were flooded, but it appears that portions of the floodplain adjacent to the river on the crest of the Monroe Uplift were not flooded (Figure 4.30). Property damage in the two Louisiana parishes (East Carrol and West Carrol) and a Mississippi County (Issaquena) in this reach of the river was $1 680 360, whereas in one county upstream (Washington), damages were in excess of $22 million and, in a downstream county (Madison, Louisiana), damages were in excess of $2.5 million (Mississippi River Flood Control Assoc., 1927). The Monroe Uplift may have substantially reduced the extent of local flooding and property damage during one of the largest Mississippi River floods. For more information on the great 1927 flood, see Barry (1997).

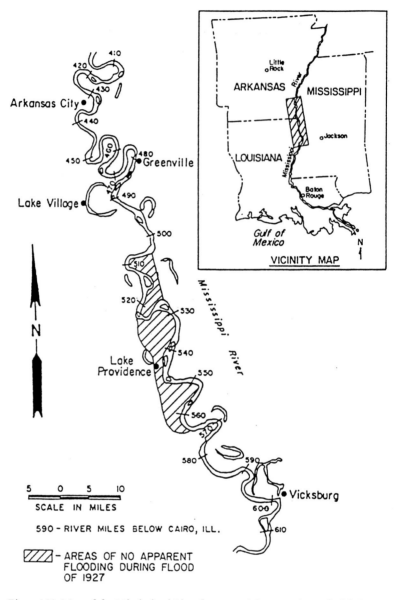

Figure 4.30 Map of the Mississippi River between Arkansas City and Vicksburg showing areas of no apparent flooding in 1927.

5

Earthquake effects

The destruction and damage of cities by earthquakes has been of prime human concern, from the 4350 BC destruction of Sodom and Gomorrah (Neev and Emery, 1995) to the twentieth-century damage to San Francisco, Los Angeles, and Tokyo. These were catastrophic events. As an example, the U.S. Geological Survey has devoted much time and money to studies in the New Madrid seismic zone in an effort to learn more about the great 1811–12 New Madrid earthquakes (see also Chapter 10), which if they reoccurred would be a national disaster.

Earthquakes not only affect cities, but they modify landforms by landsliding (Jibson and Keefer, 1989), and they can significantly affect river morphology and hydrology temporarily or permanently by triggering an avulsion. For example, the March 27, 1964 earthquake in Alaska caused regional tilting that temporarily reduced the flow of the Copper, Kenai, and Kasilof Rivers, the flow of which was opposite to the direction of tilt (Plafker, 1969). The Copper River reportedly stopped flowing for several days, and the Kenai River temporarily reversed direction and flowed back towards Kenai Lake. In addition, the tilting reduced the gradient of the Copper River by 30 cm over 7.7 km and the Kenai River by 30 cm over 16 km .

The 1811–12 New Madrid earthquakes and their effects

Between December 16, 1811, and March 15, 1812, a total of 203 damaging earthquakes occurred in the New Madrid seismic zone (Figure 5.1). Between December 1811 and February 1812, three earthquakes exceeded magnitude 8. The three major shocks occurred on December 16, 1811, January 23, 1812, and February 7, 1812 (Fuller, 1912). Johnston (1982) estimated the magnitude of the first quake at 8.6. The damage from the first event extended over a radius of approximately 700 km from the epicenter (Nuttli, 1982). A

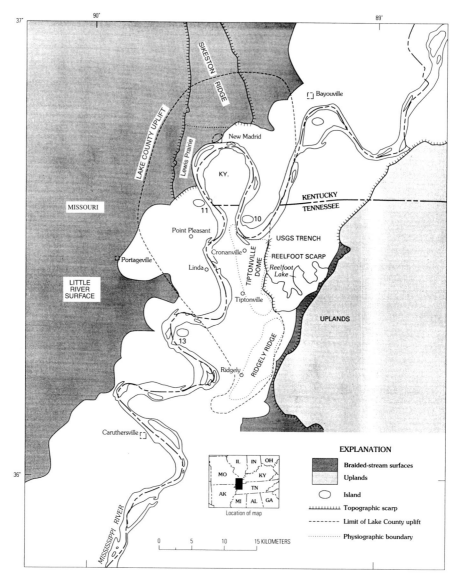

Figure 5.1 Map of New Madrid region showing geographic and geomorphic features including relationship of Mississippi River planform to the Lake County uplift (from Russ, 1982).

series of aftershocks, some damaging, followed the first event. The second major earthquake occurred on January 23, 1812. The magnitude of this event is estimated to be 8.4 (Johnston, 1982). The epicenter was probably located between Tiptonville, Tennessee and Portageville, Missouri (Figure 5.1). The last shock was the largest, occurring on February 7, 1812. Johnston (1982) assigned the earthquake a magnitude of 8.7. The epicenter was near New

Madrid, Missouri, which was completely destroyed. Fortunately in 1811 and 1812 the Mississippi valley was sparsely populated and therefore, property damage and loss of life was not great. If a similar earthquake occurred today, damage, loss of life, and injuries would make the event a national disaster; hence, the great interest in the area (Odum *et al.*, 1998).

The effects of the New Madrid earthquakes on the Mississippi River were described by several eyewitnesses because boat traffic on the river was heavy during the winter of 1811–12. Summaries of these accounts are given by Fuller (1912) and Penick (1981). For example, in a letter dated February 18, 1812, J. Smith described how the river had "wholly changed" between New Madrid and Memphis. In this reach, the development of bars and lodging of snags created severe hazards to navigation (Penick, 1981).

There is a recurring description of two distinct "rapids" in the Mississippi River near New Madrid. Penick (1981) concludes that they formed during the February 7, 1812 quake with its epicenter near New Madrid. Fuller (1912, p. 58) wrote "Faulting, at least in the surficial deposits, was not a common nor a characteristic feature of the New Madrid earthquake, although it occurred at a number of points. Some of the faults crossed the Mississippi, causing rapids and even waterfalls." William Shaler stated that a few miles above New Madrid "he came to a most terrific fall, which, he thinks, was at least 6 ft perpendicular, extending across the river . . . Another fall was formed about 8 miles below the town, similar to the one above, the roaring of which he could distinctly hear at New Madrid. He waited five days for the fall to wear away" (Fuller, 1912, p. 58). Approximately 7 miles downstream of New Madrid, Russ (1982) demonstrated an oversteepening of the natural levee profile that indicates approximately 2 m of relative uplift.

Another consistently described effect of the earthquakes on the Mississippi River was extensive bank failure. Fuller (1912, p. 89) recorded the observations of the English naturalist Bradbury, who stated "immediately the perpendicular surfaces (banks) both above and below us, began to fall into the river in such vast masses as nearly to sink our boat by the swell they occasioned." Bank erosion was also described by Lloyd (Fuller, 1912, p. 89): "At times the waters of the Mississippi were seen to rise up like a wall in the middle of the stream and suddenly rolling back would beat against either bank with terrific force . . . during the various shocks the banks of the Mississippi caved in by whole acres at a time."

Eyewitness accounts of the event commonly include references to the river "running backward" during the first series of shocks. Penick (1981) attributed this phenomenon to catastrophic bank failure and consequent generation of huge swells in the river. A rapid rise and fall of the river stage of between 1 and

4 m was described as well. Flow velocities of 8 to 13 km per hour were reported to have accompanied this change in stage (Penick, 1981). Fuller (1912, p. 90) quotes a statement that describes the heaving of the bottom of the Mississippi as "sufficient to turn back the waters." "A bursting of the earth just below the village of New Madrid arrested the mighty stream in its course." Large whirlpools, or "sucks", were reported in the vicinity of New Madrid on February 7, 1812. These features were probably generated by sand blows in the bed of the river. As the bed was disturbed, tremendous numbers of trees that were buried by sand were dislodged and floated to the surface.

Numerous islands within the channel were reported to have sunk during the earthquakes. Fuller (1912, p. 10) writes of the December 16, 1811, earthquake in the following manner: "On the Mississippi, great waves were created, which overwhelmed many boats and washed others high upon the shore, the return current breaking off thousands of trees and carrying them out into the river. High banks caved and were precipitated into the river, sand bars and points of islands gave way, and whole islands disappeared."

Both uplift and relative subsidence occurred on the Mississippi River floodplain during the New Madrid earthquakes. One of the most striking effects of the 1811 and 1812 earthquakes was the blockage of Reelfoot Creek and formation of Reelfoot Lake (Figure 5.1). The lake is located just east of the Tiptonville dome and Reelfoot scarp in northwestern Tennessee, and it occupies several abandoned channels of the Mississippi River, which were probably oxbow lakes at the time of the earthquakes (Russ, 1982).

The most significant surface deformation associated with faulting in the area is the Reelfoot scarp. Russ (1982) concluded that most of the deformation on the Tiptonville dome took place during the last 2000 years. Furthermore, the uplift occurred primarily during at least three distinct earthquakes (two pre-1800, one during 1811–12). Reelfoot Lake is one of the "sunklands" of the lower Mississippi Valley. Several other sunklands are present on the west side of the river along the St. Francis and Little Rivers. The formation of the lake suggests that Tiptonville dome was uplifted, blocking Reelfoot Creek, or the area where the lake presently exists subsided or was compacted during the event.

The Lake County uplift is the most prominent surficial expression of modern deformation in the New Madrid seismic zone (Figure 5.1). Russ (1982) constructed topographic profiles of natural levees and abandoned channels of the Mississippi River, as well as profiles across the Reelfoot scarp, and demonstrated that the uplift is of tectonic origin. The uplift forms an elongate topographic high that rises as much as 10 m above the surrounding Mississippi River floodplain between Ridgely, Tennessee and just north of

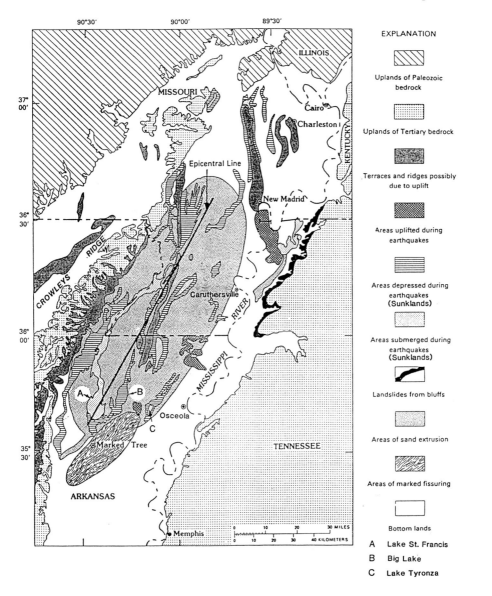

Figure 5.2 Earthquake features of the New Madrid district (modified from Obermeier, 1989).

New Madrid, Missouri, and it consists of two main centers, Tiptonville dome and Ridgely ridge (Figure 5.1).

Another area that can be attributed to recent surface deformation is the St. Francis sunklands located between Osceola, Arkansas, and Crowleys Ridge (Figure 5.2). The sunklands include Big Lake and St. Francis Lake and other low-lying swampy areas along the St. Francis and Little Rivers. Guccione *et al.*

(1993) and Guccione and Hehr (1991), using topographic, geomorphic, sedimentologic, archaeologic, and historic data, determined that the present Big Lake was formed during the 1811–12 earthquakes. Buried soils and fluvial deposits with intervening lacustrine deposits suggest that the present Lake St. Francis was also formed by the 1811–12 earthquakes, and it may have existed at two other times in the past 8000 years (Guccione and Van Arsdale, 1994). Guccione et al. (1994) and Guccione and Van Arsdale (1994) estimate that a total of 2.5 to 3.2 m of Holocene subsidence has occurred in this area.

Other earthquake-generated features within the study area that have been examined in considerable detail (Figure 5.2) include sand blows (Fuller, 1912; Saucier, 1977, 1989; Heyl and McKeown, 1978; Obermeier, 1989), shallow subsurface liquefaction deposits (Russ, 1982; Saucier, 1977, 1989); fissuring (Fuller, 1912), and landslides (Fuller, 1912; Jibson and Keefer, 1993).

It is obvious that the New Madrid earthquakes had temporary effects on the Mississippi River, but even after almost 200 years, there is evidence that the effects persist. Any attempt to relate the earthquakes to channel-bed profiles of such a large river can be extremely difficult due to large natural variations in pool and crossing elevations both spatially and through time. In contrast, water-surface profiles provide relatively smooth reference planes that broadly reflect both channel geometry and slope. In order to maintain a database regarding low flow conditions for navigation purposes, the U.S. Army Corps of Engineers surveys the water surface every few years. For example, bankfull and medium stages were obtained from surveys dated 1880 and 1915 from Columbus, Kentucky, to Caruthersville, Missouri (Figure 5.3). The profiles, which cross the Lake County uplift, show a distinct profile convexity at both bankfull and medium stage relative to linear best-fit lines in 1880 and 1915. This convexity extends from approximately river mile (RM) 870 to RM 910, which coincides with the Lake County uplift. River and valley miles will be used in all discussions of the Mississippi River as all locations (navigation, lights, etc.) are located by river mile (RM).

Surveyed low-water profiles for 1880, 1976, and 1988, which reflect average bed configuration, reveal that the more recent low-water profiles are below the 1880 profile, which indicates that the river has degraded (Figure 5.4). The broad low-water convexity at New Madrid, which is present in the 1880 profile, is not present on the 1976 and 1988 profiles, which in contrast have a distinct stepped appearance (Figure 5.4). The steepest section of the recent water-surface profiles (1976 and 1988) is located just downstream of New Madrid, Missouri, at about RM 888. At this location, the 1988 profile shows a water-surface drop of approximately 1.2 m over approximately 3.2 km. This relatively steep water surface may reflect the exhumation of a

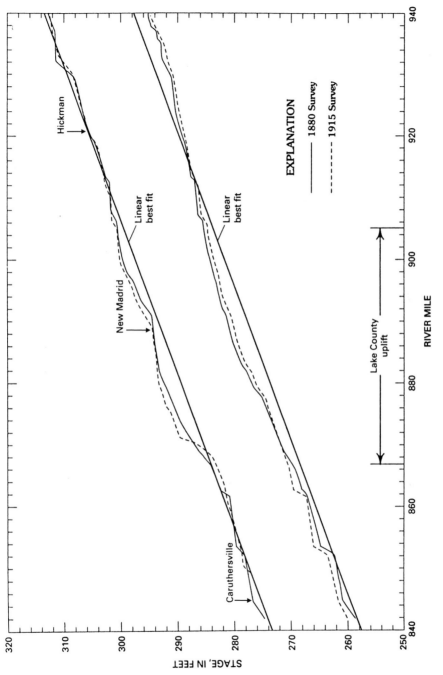

Figure 5.3 Bankfull (upper lines) and medium stages of Mississippi River and location of Lake County uplift as defined by Russ (1982).

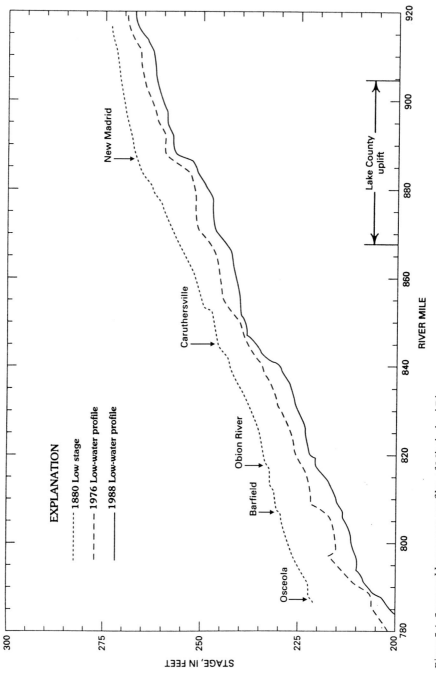

Figure 5.4 Surveyed low-water profiles of Mississippi River.

fault scarp in the Mississippi River channel. The steep channel section is located approximately 8 km upstream of a fault mapped by O'Connell *et al.* (1982), where a waterfall formed following the February 7, 1812, earthquake according to eyewitness accounts (Johnston, 1982).

Any vertical offset of the channel bed during the earthquake would have generated turbulence and bank erosion, but the fault scarp would have been eroded rapidly, and eyewitness accounts indicate that the rapids were short-lived. The eroded scarp probably persisted as an anomalously steep channel section that lengthened and flattened through time. Subsequent erosion of the bed would cause upstream migration of the steep channel section. Although the vertical exaggeration of Figure 5.4 produces abrupt scarp-like features, note that the steep sections extend over several miles of channel.

The water-surface profiles and the historical accounts suggest that, during the 1812 event, vertical offset on the cross-rift trend produced convex water-surface profiles over the structure. Since the earthquake, the river has eroded its bed upstream from the original location of the fault (RM 883) to approximately RM 888, and, consequently, it has reduced the convexity.

A second point of slight water-surface steepening near RM 870 (Figure 5.4) occurs where the Mississippi flows off of the Lake County uplift (Russ, 1982). No fault has been mapped at this location to date; however, the water-surface profiles suggest that the western margin of the Lake County uplift may be bound by a fault or monoclinal fold. A large clay plug intersects the channel at this location (the Tiptonville meander; Russ, 1982); therefore, the steep water surface may be due to resistant clay, which limits bed degradation and constitutes a local high in the channel at that location. It is also quite possible that Tertiary-age sediments are exposed in the channel at this location. Relatively high water-surface slopes at Caruthersville, Missouri (RM 848) are probably also the result of lithologic controls by a clay plug and possibly by the presence of Tertiary-age sediments.

Eye witness accounts of the effects of the earthquakes are consistent in describing tremendous numbers of bank failures that introduced large quantities of sediment into the river. Walters (1975) and Walters and Simons (1984) assumed that this great increase in sediment load could affect the river far downstream, and he notes that bank caving was described from the Ohio River almost to Memphis, Tennessee. Using a map prepared by three Army officers in 1821, Walters determined that between Cairo, Illinois and the mouth of Red River between Natchez and Baton Rouge, Louisiana, the average channel width was 3100 feet (Figure 5.5). From about 40 miles upstream of Memphis (30 miles downstream of Osceola, Arkansas) channel width was well above average, whereas downstream it was about or below average.

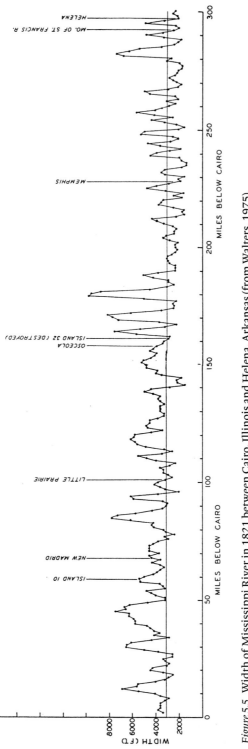

Figure 5.5 Width of Mississippi River in 1821 between Cairo, Illinois and Helena, Arkansas (from Walters, 1975).

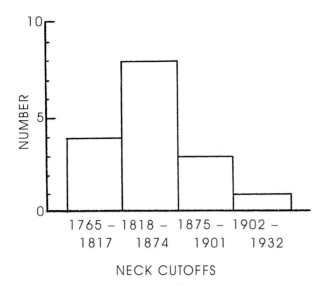

Figure 5.6 Number of meander cutoffs for four periods (from Walters, 1975).

In addition, the number of cutoffs increased after the earthquake (Figure 5.6) from four in the 52-year period, 1765 to 1817, to eight in the 56-year period between 1818 to 1874, after the earthquakes. The Mississippi River example provides a clear and convincing case of drastic and long-term effects of earthquakes on alluvial rivers.

Avulsion

Large alluvial rivers that flow down wide alluvial plains can avulse as a result of faulting and earthquakes. For example, the Indus River in Pakistan is flowing on an alluvial ridge of its own making. The river has avulsed widely in the past (Figure 7.17), and it is probable that earthquakes have played a role. For example, there are numerous active faults in the Indus Valley (Kazmi, 1979). The most spectacular effect of active faulting is in the Rann of Cutch fault zone. In this region of the lower Indus Valley in 1819, a severe earthquake resulted in the uplift of a 16 km wide and 80 km long tract of alluvial land with a relief of about 6 m. This feature was locally known as Alah Bund (Oldham, 1926). The eastern branch of the Indus was blocked by the formation of the Alah Bund (Oldham, 1926). The channel at that time was dry, but flow was re-established during a flood in 1828. For discussion of recent history of the Indus, and its impact on an ancient (Harappan) civilization, see Holmes (1968), Jorgensen *et al.* (1993), and Flam (1993).

The Brahmaputra River in Bangladesh avulsed 80 to 100 km to the west in

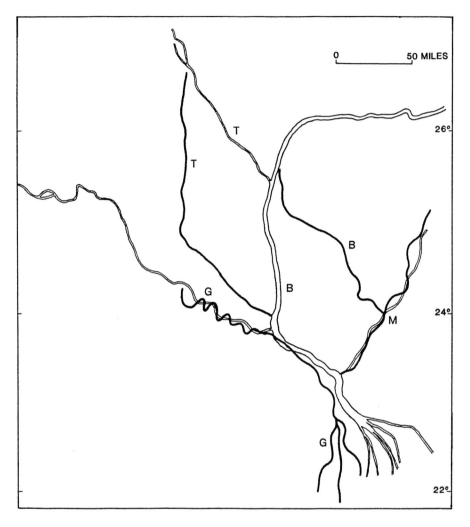

Figure 5.7 Sketch map showing positions of Brahamaputra River (B), Megna River (M), Tista River (T), and Ganges River (G) in 1779 (solid lines) and at present (double lines).

the late 18th century during a time of large earthquakes (Figure 5.7). The earthquakes may have played a role in this avulsion, but tilting (Morgan and McIntire, 1959; Johnson and Alam, 1991) and subsidence or simply the selection of a more direct route to the sea could have caused the avulsion. Nevertheless, the bank failures along the Mississippi River suggest that major failures that result in a loss of natural levees could contribute during high water to avulsion. Liquefication of bank sediments, sand blows, and fractures all could contribute to this process. For example, failure of the human-constructed levees of the Yellow River, the bed of which is perched 10 m above the surface of the North China Plain, would obviously cause avulsion, in addition to a major human tragedy (Zhou and Pan, 1994).

Discussion

The effect of major earthquakes on cities is widely known and a repetition of the New Madrid earthquakes would cause great damage in St. Louis and Memphis in addition to numerous small communities located in the alluvial valley. In Bangladesh, there is a French plan to limit annual flooding by the construction of levees (embankments) with an average height of 4.5 m and a maximum height of 7.4 m. This plan is designed for a country where existing embankments are neglected. The levees would be constructed in one of the world's most earthquake-prone regions where "the largest earthquake on land known to seismologists, registering 8.7 on the Richter Scale" occurred in 1897 (Boyce, 1990). Boyce (1990) attributed the previously discussed avulsion of the Brahmaputra River to earthquakes. In areas of alluvial sediments with high water tables such as Bangladesh, earthquakes lead to liquefaction, which causes sediments to behave like viscous fluids. The glossy summary of the French plan that was distributed to political leaders and the public does not mention earthquakes, although failure of the embankments would expose people to sudden floods of catastrophic magnitudes.

Part III

Effects of unmeasured deformation

6

Response of channel morphology, sedimentology, hydrology, and hydraulics

In the previous discussions the response of rivers to active tectonic deformation has focused on changes in the longitudinal profile, aggradation, degradation, and channel planform or pattern, whereas cross-section morphology, flooding, hydraulic character of the channel, and changes of sediment characteristics have not because the former can be well documented at the scale of map and aerial photography, whereas the latter requires field surveys. Nevertheless, the deformation of rivers will affect sediments deposited by the river, and changes in channel pattern caused by deformation will appear as changes in sedimentary facies. Therefore, detailed studies of modern conditions should make it possible to detect facies changes of sediments from known tectonic movements (Slack, 1981). In addition, the variability of overbank flooding along a river that is subject to tectonic activity (Figure 4.30) should be of potential concern to hydrologists, riparian landowners, and insurance companies. For example, field work was conducted on Bogue Homo and Tallahala Creek (see Chapter 4) after a period of high flow. Due to a period of heavy rainfall, flows peaked approximately three days prior to the commencement of data collection. As a result, a distinct high-water mark was visible, and it was surveyed at each cross-section where its presence was identifiable. The location of the high-water mark provides valuable information concerning the distribution of overbank flow as related to active tectonics (Figures 6.1 and 6.2). The spatial relationship between the high-water mark and top bank is very similar for both streams. For example, during the flood, flow was overbank upstream of the uplift axis, but it was well within the channel at and downstream of the uplift axis. The change of flood characteristics at the uplift axis is striking. It is the anomalously deep channel at and below the uplift axis, a result of degradation due to uplift, that has confined the flow and prevented flooding downstream.

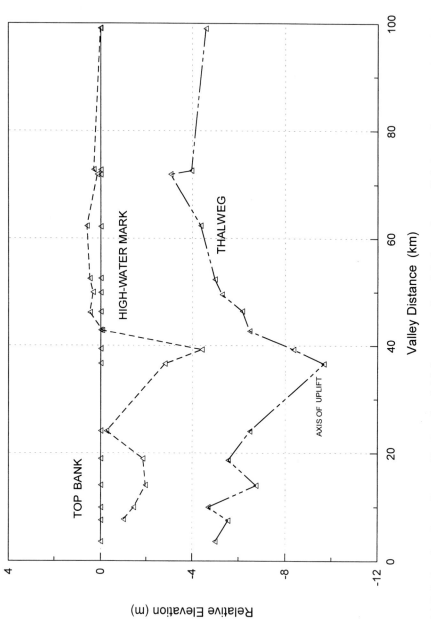

Figure 6.1 Relative elevations of normalized top bank, high-water mark, and thalweg along Tallahala Creek. Upstream is to the right.

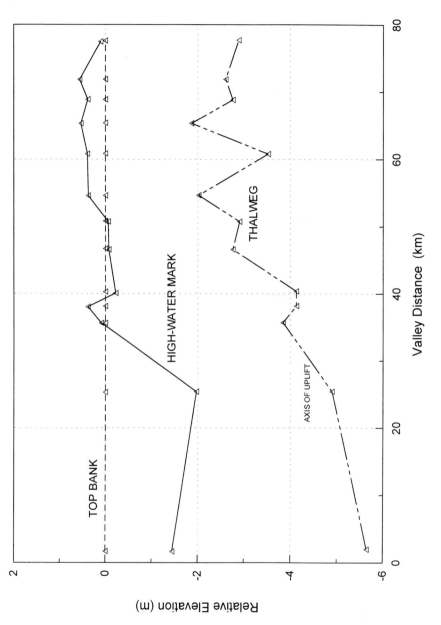

Figure 6.2 Relative elevations of normalized top bank, high-water mark, and thalweg along Bogue Homo. Upstream is to the right.

Another example is provided by the Compton Landing Gas Field, which underlies the Sacramento River upstream of Sacramento, California. The gas field is related to an active uplift, which has its long axis perpendicular to the river (L.H. Lattman personal communication). The uplift has warped the valley profile, and slope has been decreased above the uplift and increased below it. The reduction in slope upstream of the uplift has reduced the velocity of the river and elevated the stage. During floods, the river frequently overtops and/or breaches its natural levees and floods the adjacent area. Breaching of the levee is clearly related to a known structure, which is deforming the present valley floor and affecting the flood hydrology and hydraulics of the river.

In view of these obviously important details of river morphology, sedimentology, and hydrology, Jorgensen (1990) extended the studies discussed in previous chapters by exploring alluvial river response to active tectonics in more detail. Specifically, his objectives were to determine the effect of active tectonics on the variability of:

1. channel size and shape,
2. channel hydrology and hydraulics, including size and frequency of bankfull flow, flow velocity, and stream power,
3. sediment characteristics of the channel bed, bars, and banks,
4. rate and character of erosion and deposition.

Channel pattern and the longitudinal profiles, of course, were a necessary part of the investigation, but the emphasis was on more local features. The response of the rivers varied because of the sedimentological, hydrological, and historical conditions that make each river unique. Four rivers flowing in valleys that have been or are being deformed by active tectonics were investigated by Jorgensen (1990) as follows:

1. Neches River, east Texas, a suspended-load, meandering stream crossing an uplift with a subsiding axial graben or fault zone,
2. Humboldt River, Nevada, a mixed-load stream crossing a zone of subsidence in the Basin and Range Province,
3. Sevier River, Utah, a gravel-bed stream crossing an uplift,
4. Jefferson River, Montana, a gravel-bed stream crossing an uplift.

Channel morphology is sensitive to many factors other than tectonic deformation. Therefore, these rivers were selected because they are not significantly influenced by changes of rock type, bedrock exposure, tributary sediment loads, or recent changes of sediment supply and water discharge. Nearby, National Geodetic Survey (NGS) surveys support the conclusion that the Neches and

Humboldt Rivers are affected by active tectonics. Evidence for active tectonics in the Sevier and Jefferson River valleys is seismic and geomorphic.

Neches River, Texas

The Neches River meanders and it is flanked by a wide, frequently inundated floodplain. The river carries a high suspended-sediment load, and it deposits fine sand and mud as channel bars and on the floodplain. It is one of several streams, which rise in and drain the coastward-dipping Mesozoic and Cenozoic sediments of the Gulf Coastal Plain (Figure 6.3). It heads in a structural subdivision of the coastal plain known as the East Texas Basin (Murray, 1961). Ongoing tectonic movement is indicated by NGS benchmark releveling and by seismicity just east of the study area. A releveling line between Tyler and Nacogdoches (Figure 6.3) indicates 130 mm of south-downthrown displacement at the eastern end of the Mt. Enterprise Fault Zone between 1920 and the mid-1950s (Collins *et al.*, 1981).

The Neches River flows transversely across a zone of uplift that is further deformed by extensional faulting at the axis of the uplift (Figure 6.4). Geomorphic expression of the uplift and faulting includes the deformation of terrace profiles, tilting of paleochannels, truncation of paleochannels, and fault or fault-line scarps. Morphologic criteria including channel shape, valley and channel slope, and floodplain character were used by Jorgensen (1990) to divide the 32 km study area into five reaches (Table 6.1), and the character of each reach is summarized in Figure 6.5. Reach 1 is the control reach, which is presumed to be unaffected by tectonism, and it is at the downstream end of the study area. Sixteen cross sections were surveyed across the channel and floodplain. Cross-section 16 is upstream of Reach 5, and it is located at the upper margin of the fault zone (Figure 6.4). Reaches 4 and 5 are in the area affected by relative subsidence. The channel in Reach 3 is incising the upthrown block or graben shoulder immediately downstream of the fault zone. The valley is narrow in Reach 3, but the channel is not deeply incised. Lastly, Reach 2 is located on the steep (Figure 6.4) downstream flank of the uplift.

The most pronounced effect of the morphologic changes is great variability in the frequency of overbank flow (Figure 6.6D), which primarily reflects the influence of channel slope and channel size (Figure 6.6A, Table 6.1) on channel capacity. In the field, an observed discharge of 55.6 m³/s following early winter thunderstorms was at or below the banks in Reach 1 and Reach 5. It was below bank level on the downstream flank of the uplift (Reach 2) and well out of bank in Reaches 3 and 4.

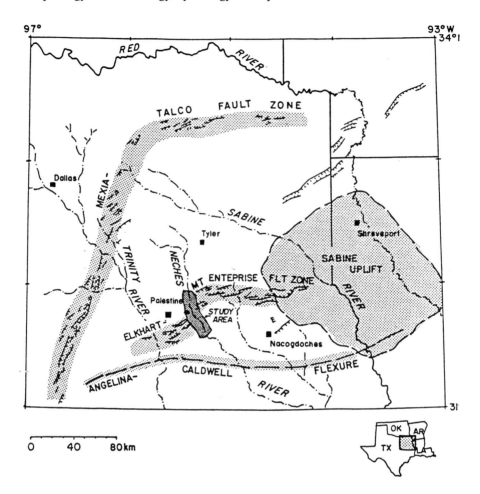

Figure 6.3 Map showing the location of the Neches River study area within the East Texas Basin and associated geologic structures (from Jorgensen, 1990). Redrawn from Baumgardner (1987).

The change of bankfull channel size is caused largely by a decrease of depth from the top of the study area through the fault zone and then a rapid increase of channel depth through Reaches 3 and 2 (Figure 6.6B, Table 6.1), with less prominent but similar changes of channel width (Figure 6.6C). As a result of these changes plus the change of channel slope (Table 6.1), the frequency of overbank flow (Figure 6.6D) increases through the deformed zone and, reaches a maximum in the narrow valley of Reach 3 (Table 6.1, Figure 6.6). In this reach, flow is overbank, on average, 93 days a year (Table 6.1). The variation in morphologic and flow characteristics (Figure 6.5) was examined at a single discharge, the 2-year flood ($180\,\text{m}^3/\text{s}$), which was selected because of the presence of the prominent high-water mark in the study area (Table 6.1).

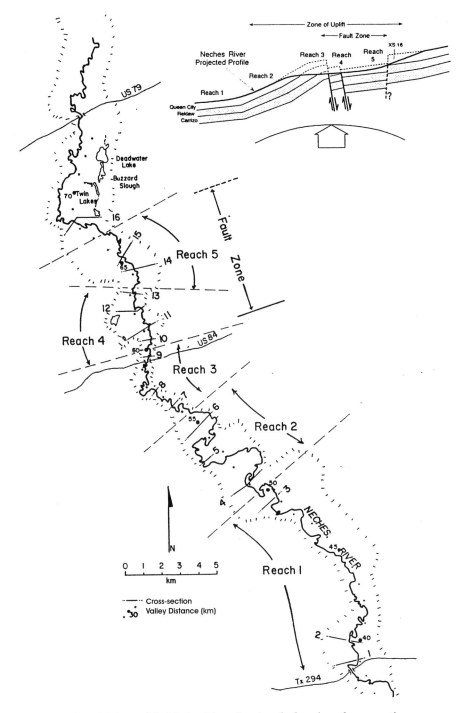

Figure 6.4 Map of the Neches River showing the location of cross-sections, reaches, and valley distance (from Jorgensen, 1990). Geologic interpretation (not to scale) compiled from information in Barnes (1968), Wood and Guevara (1981), and Jackson (1982). Flow is to the south.

Table 6.1. *Quantitative summary of Neches River response to deformation*

Reach		1	2	3	4	5	XS16
Sinuosity		1.95	2.72	1.93	1.65	1.75	1.9
Valley slope		0.00032	0.00078	0.00032	0.00017	0.00023	0.00059
Bankfull Channel characteristics							
Area	(m²)	115	124	64	35	65	98
Discharge	(m³/s)	79	102	25	68	54	94
Flow exceedence	(days/yr)	31	26	93	38	50	13
Width–depth ratio		9.6	6.7	12.1	11.0	9.5	8.7
Width	(m)	46	44	41	31	34	49
Depth	(m)	4.8	6.6	3.4	2.8	3.6	5.7
Flow characteristics of the 2-year flood							
Water surface slope		0.00020	0.00022	0.00012	0.00012	0.00022	0.00025
Width	(m)	663	642	483	1189	995	858
Thalweg depth	(m)	5.5	7.2	5.1	3.4	4.0	6.0
Average velocity	(m/s)	0.69	0.63	0.36	0.34	0.45	0.81
Channel shear stress	(N/m²)	5.4	7.3	4.1	1.9	4.9	5.0
Unit stream power	(W/m²)	3.6	4.7	1.7	0.6	2.3	4.1
Sediment characteristics							
Median floodplain size	(mm)	0.06	0.06	0.04	0.06	0.06	0.09
Median bar size	(mm)	0.11	0.14	0.09	0.12	0.06	0.19
Median bed size	(mm)	0.21	0.17	0.16	0.13	0.08	0.18

The combination of these hydrologic and morphologic responses reduce stream power in the fault zone and increase it in the incising reaches (Table 6.1). Total stream power per unit channel length (Ω) is one measure of sediment transport capacity, and it is defined as $\Omega = \gamma QS$, where γ = unit weight of water (9810 N/m³), Q = channel discharge, and S = energy gradient, which is frequently approximated by water surface slope (Bagnold, 1980). Stream power expended per unit area of channel bed, $\omega = \Omega/w$ where w = channel width, is used in calculations of sediment transport (Bagnold, 1980). Unit stream power was calculated for the in-channel portion of the 2-year flood discharge (Table 6.1). Clearly, the Neches River has the least flow energy in the fault zone, where it is capable of transporting less sediment.

In general, the sedimentologic response of the Neches River is expected (Table 6.1); the river is sluggish in the fault zone on the upstream portion of the uplift (Reaches 4 and 5), and incision is occurring in the reaches of relative uplift (2 and 3). Deposition or simply the slackening of incision in response to the lowered valley slope in the fault zone has caused finer sediment to be deposited in the channel (Table 6.1), which in turn has resulted in reduced channel carrying capacity, frequent flooding, and low stream power.

Downstream ←————————————————— Upstream

	Reach 1	Reach 2	Reach 3	Reach 4	Reach 5	XS 16
Position	control	downstream flank of uplift	fault-zone margin and uplift apex	downstream part of fault zone	upstream part of fault zone	upstream margin of fault zone
Vertical Change	stasis?	incision	incision	aggradation	aggradation	incision
Cross Section shape		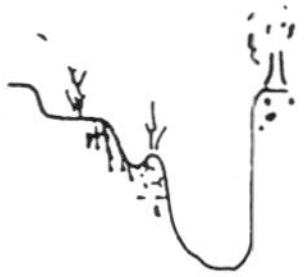			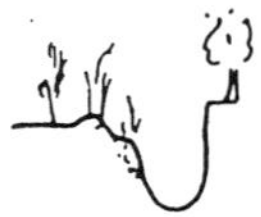	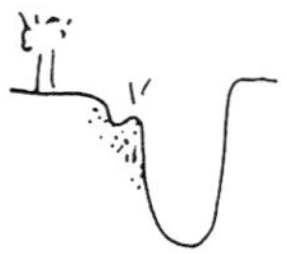
Valley Geometry	channel bordered by high natural levees and lower elevation flood basins	high, low relief floodplain and low terrace surfaces, sand deposition occurs as splays on floodplain	narrow U-shaped channel bordered by abandoned channels (left bank, xs 8) and terraces	very wide, often swampy, floodplain with frequent active and inactive sloughs	low levees, floodplain has few well-developed sloughs but large, shallow flood basins	high, infrequently-flooded floodplain surface characterized by few sloughs or flood basins.
Channel and Water-Surface Profiles	comparable water-surface, bed and bank profiles; 2-year floodline at levee top	steep valley slope but channel slope similar to Reach 1, irregular bank profile due to presence of multiple erosional, surfaces	2-year floodline very high above floodplain surface, bed slopes more than bank or water surface	2-year floodline not high above floodplain, but demonstrates extensive flooding, bed slope nil but bank slope very high	low bed slope with 2-year floodline just at level of banks, bank and water-surface slope similar to Reach 1	may be affected by dam upstream, 2-year floodline within channel, steep, deep thalweg profile
Bank and Sediment Characteristics	steep cutbanks and muddy-sand, low relief channel-margin bars	channel deep with steep, actively-eroding cutbank, well-sorted fine sand, high relief point bars	bank elevation very low but sandy and flanked by elongate abandoned bars on adjacent floodplain	low, muddy bars at alternate channel margins, few cutbanks and abundant channel bounding vegetation	low, muddy channel bars and occasional sandy cutbank, mud and channel vegetation increase downstream	high, resistant, sandy cutbank and high relief, sandy point bars.

Figure 6.5 Geomorphic characteristics of study reaches of the Neches River (from Jorgensen, 1990).

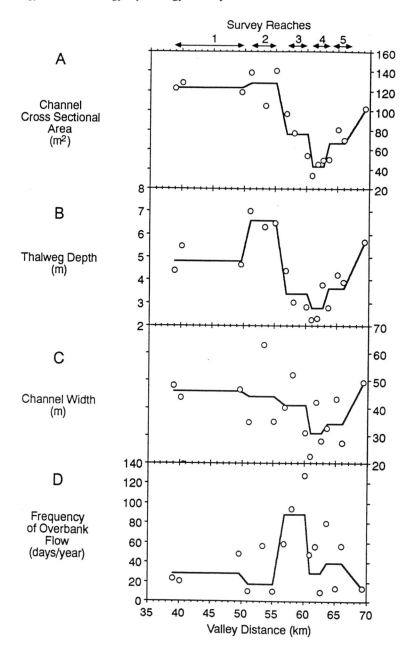

Figure 6.6 Morphologic and hydrologic characteristics of the bankfull channel. The solid line connects reach averages (from Jorgensen, 1990).

In summary, the Neches River has apparently been unable to erode and deposit enough sediment to maintain a smooth profile across the Elkhart Graben–Mt. Enterprise Fault Zone. The resulting valley slope and change from aggradation to incision have affected channel morphology, flow characteristics, and the style and grain-size of channel and floodplain sedimentation (Table 6.1, Figure 6.5). Reduction of stream power and valley slope by subsidence and/or backtilting in the fault zone (Reaches 4 and 5) has resulted in a reduction of channel size, deposition of increasingly finer-grained sediment on a wide floodplain, generally finer-grained sediment on channel bars, and decreased meander activity. Both the shape of the channel and the amount and size of sediment deposited on channel banks is a reflection of the tectonic control on channel slope and sediment transport.

Humboldt River, Nevada

The Humboldt River is the largest river in the United States that does not ultimately reach the sea, and it is one of only a few rivers that cross the Basin and Range Province from east to west (Figure 6.7A). Jorgensen (1990) studied four reaches of the river (Figure 6.8, Table 6.2), that are located 20 km downstream of Winnemucca in a segment of the river referred to as Rose Creek Narrows (Hawley and Wilson, 1965). Here the Humboldt River flows around the northern terminus of the East Range, which is bounded to the west by a west-dipping normal fault. A summary of the characteristics of each reach is presented in Table 6.2.

Evidence of ongoing tectonic activity is provided by regional seismicity and a NGS releveling line, which indicates subsidence of the Rose Creek area (Figure 6.7B) relative to Winnemucca and Mill City. The data are of poor quality because some stations were established in alluvium, but they indicate relative downwarp of the area near Rose Creek. Subsidence of the Rose Creek area is also supported by the fact that a basalt flow, which crops out just north of the river in the study area (Little Tabletop Mountain, Figure 6.8), has been faulted down beneath valley alluvium.

Reach 1 is downstream of the zone of subsidence, and Reach 4 is at the upstream limit of the subsiding zone. Channel shape changes in several ways as the Humboldt River passes through the Rose Creek Narrows, Reaches 2 and 3. There is a decrease of channel sinuosity in this zone of subsidence and an increase in the size of the bankfull channel (Table 6.2, Figure 6.9).

Width–depth ratio, width and discharge for the bankfull channel increase through the zone of subsidence (Figure 6.10), primarily due to changes of width rather than bankfull depth. Although bankfull discharge increases,

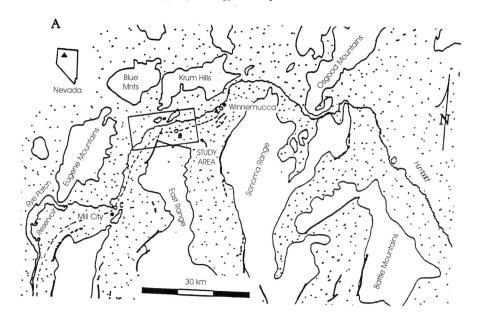

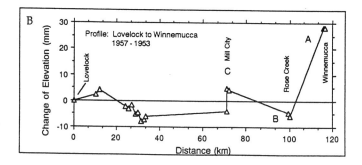

Figure 6.7 Map showing the location of the Humboldt River study area (A). Redrawn from Stewart (1978). River flow is to the west. Heavy lines on A are mapped faults in alluvium and range-front faults. Stippled areas are underlain by Quaternary alluvium. Dashed line near the river southwest of Winnemucca is approximate location of releveling line (B) (from Jorgensen, 1990).

unit stream power at bankfull flow decreases (Figure 6.10E). Unit stream power is greatest in the steep, sinuous Reach 4 (Table 6.9).

Water surface slope, average depth, and velocity in Reaches 2 and 3 are all lower than in Reach 4 (Table 6.2). The low velocity of Reach 1, despite the greater channel depth and water surface slope, is caused by a greater channel roughness caused by bed irregularity. Low flow velocity in the middle reaches

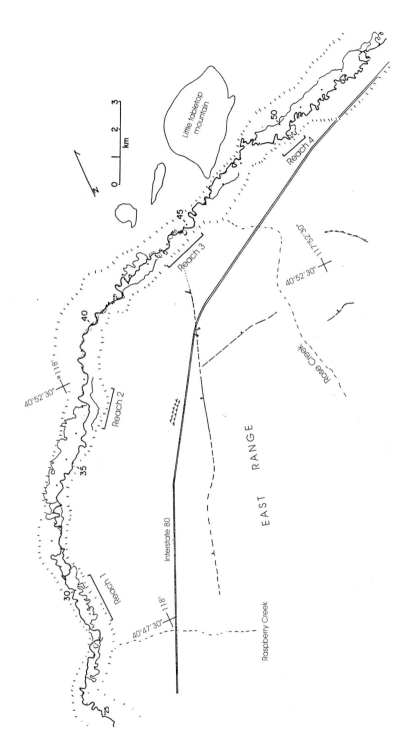

Figure 6.8 Map of the Humboldt River planform, redrawn from 1:24 000 topographic maps and orthophoto quadrangles and field mapping (from Jorgensen, 1990).

Table 6.2. *Quantitative summary of Humboldt River response to deformation*

Reach		1	2	3	4
Sinuosity		1.45	1.3	1.3	2.1
Valley slope		0.00054	0.00052	0.00046	0.00084
Bankfull channel characteristics					
Area	(m²)	134	165	277	100
Discharge	(m³/s)	109	142	224	124
Width–depth ratio		15	28	32	14
Width	(m)	59.4	95.4	124.5	48.7
Depth	(m)	3.86	3.41	3.91	3.48
Flow characteristics of the mean annual flood					
Water surface slope		0.00063	0.00048	0.00047	0.00057
Width	(m)	89	74	97	46
Thalweg depth	(m)	3.53	2.73	2.69	3.18
Average velocity	(m/s)	0.82	0.88	0.75	1.12
Average shear stress	(N/m²)	8.3	7.0	6.3	10.3
Unit stream power	(W/m²)	11.3	6.9	5.4	12.2
Sediment characteristics					
Median bar	(mm)	1.9	2.2	2.4	3.2
Median bed	(mm)	0.7	1.2	1.5	1.7
Median bank	(mm)	0.4	0.8	1.0	1.4

is primarily a reflection of the flow irregularity caused by the large bars (Reach 3) and low slope (Reaches 2 and 3). The result of these flow changes is that unit stream power and average boundary shear stress (Table 6.2) decrease in the Rose Creek disturbance relative to upstream and downstream.

There is a relatively rapid downstream fining of bed, bar, and bank sediments through the study area (Table 6.2). Despite the availability of coarse material in the underlying sediments, the median size of the sediment sampled from riffles ("bed"), and the median bar sediment size decrease (Table 6.2). The downstream reduction of grain size suggests that net deposition of sediment is occurring, which reflects the selective transport of aggrading streams (Mackin, 1948; Dawson, 1988; Paola, 1988). For example, rapid downstream decrease of boundary shear stress, coupled with high rates of channel shifting along the Green River, produces a rapid downstream grain size diminution (Dunne, 1988).

In summary, channel morphology and process clearly change in the Rose Creek Narrows. Stream power and the capacity of the channel to transport sediment are reduced (Table 6.2), and aggradation occurs. Through the zone of subsidence and aggradation, the channel is less sinuous (Table 6.2), but it has a water surface slope that is lower than the more meandering segment of

Downstream ————————————————→ Upstream

	Reach 1	Reach 2	Reach 3	Reach 4
Position	control, assumed to be below zone of subsidence	below central zone of subsidence, regain long-term incision(?)	zone of subsidence	upstream limit of zone of subsidence, forward tilt
Vertical Change	stasis	incision(?)	aggradation	d/s tilt, incision(?)
Channel Pattern				
Channel Change	slow lateral migration of channel and bars, slow extension and translation of infrequent large meander bends	downstream half: meander initiation and growth, upstream half: downstream shift of alternate bars	rapid meander growth and frequent cutoff, chaotic channel avulsion and reoccupation of former channels	slow lateral migration and growth of very tight meander bends, simultaneous abandonment of several bends
Bar Characteristics	narrow, high-relief, alternate bars that fine and migrate downstream gravelly-sand, erosional at upstream margin	u/s half: very wide, low-relief sand and gravel alternate bars with marginal channels, d/s half: bank-attached bars.	wide, high-relief, scroll bars where channel not recently changed, very large chaotic bars near neck and chute cutoffs	narrow, high-relief, well sorted, gravel bars in a narrow slightly sinuous channel, larger coarse point bars in meanders
Profile Sequence	deep scour holes at bend apex and near frequent, resistant, sandy-clay, channel-fill beds	less irregular, wider channel with flatter pools and riffles than Reach 1	steep, gravelly riffles near recent channel changes and occasional, deep, dune-covered pools	narrow, deep pools, wide, smooth riffles, occasional steep reach near avulsions
Bank and Sediment Characteristics	high, predominantly sand to sandy-clay lateral accretion and channel-fill deposits, thick fine-grained overbank layer	low-relief banks with gravelly sand, cross-stratified bar deposits and thin silty overbank layer	frequent, several-meter-high sets of tabular cross stratified bar deposits in high, steep banks, some without fine cap	low, vertical, fine-grained banks composed of gravelly-sand to muddy-sand lateral-accretion beds
Valley Position	high smooth floodplain with single, distinct abandoned channel, frequent terrace remnants	u/s half: low, smooth floodplain, d/s half: high, smooth floodplain and recently abandoned slough, low terraces	narrow, highly irregular floodplain with several stages of abandoned channels and meanders	low, very wide, flat floodplain with distinct abandoned channels and infrequent low terrace remnants

Figure 6.9 Geomorphic characteristics of study reaches of the Humboldt River (from Jorgensen, 1990).

Reach 4. The change of channel morphology, channel pattern, plus rate and style of channel change (Table 6.2, Figure 6.9) is reflected in the sedimentology; bar deposits change from lateral-accretion deposits to planar foresets in the zone of deposition. Lastly, the deposition in this segment of the Humboldt River causes rapid downstream sediment fining (Table 6.2).

Sevier River, Utah

The Sevier River in south-central Utah flows down the axis of one of a series of extensional valleys cut into the eastern margin of the Colorado Plateau (Figure 6.11). It is a gravel-bed river that is being affected by tectonism associated with extension in the Intermountain Seismic Belt (Smith and Sbar, 1974). The valley, which was originally formed by extension, is now

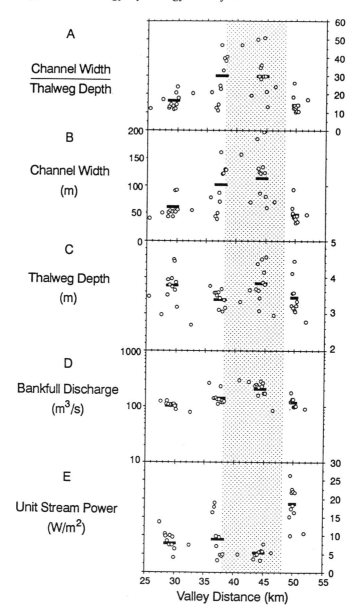

Figure 6.10 Morphologic and hydraulic characteristics of the Humboldt River bankfull channel. Solid lines are reach averages and stippled area marks the proposed zone of subsidence (from Jorgensen, 1990).

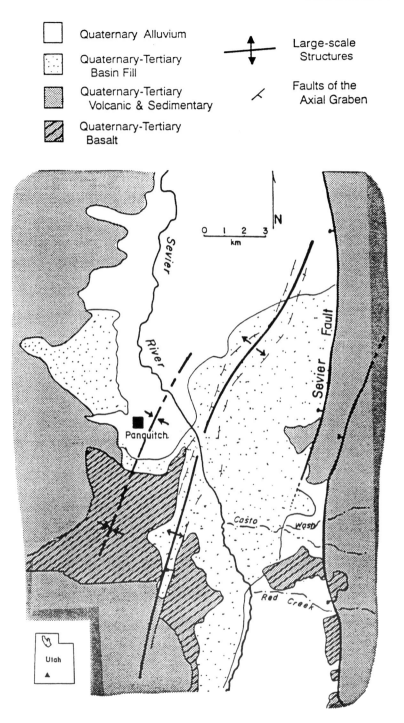

Figure 6.11 Geologic map of the Sevier River study area. Geology from Carpenter and others (1967). Sevier River flows to the north (from Jorgensen, 1990).

being re-deformed, and one segment of the valley is being uplifted relative to other segments. These structural valleys disrupt the relatively flat-lying Mesozoic- and Cenozoic-age volcanic and sedimentary rocks of the high plateaus. The Sevier Valley near Panguitch (Figure 6.12) is a half-graben that is bounded on the east by the Sevier Fault, which is a west-dipping normal fault extending from northern Arizona to northeast of Panguitch. Displacement on the Sevier Fault is up to 900 m (Lundin, 1989), primarily as dip-slip.

The Sevier River cuts obliquely across a middle to late Quaternary uplift with a small axial graben (Figure 6.12). This uplift has deformed alluvial surfaces that slope smoothly toward the river upstream and downstream of the deformed zone. The Sevier River has incised older valley fill deposits through the uplift, and it has become steeper and straighter in the downstream and middle portions of the uplift. In the upstream portion of the uplift, the river is less entrenched, the valley slope is gentle, and the channel is meandering (Table 6.3, Figure 6.13). Deposition occurs downstream of the uplift, where the channel crosses a syncline, which may also be active. Whether because of the syncline or simply the deposition of sediment eroded from uplift, older alluvial surfaces are buried by more recent alluvium in this segment. Although no NGS surveys are available, ongoing tectonic activity is indicated by deformation of alluvial surfaces and by five historic earthquakes on the axis of the faulted anticline.

Jorgensen (1990; Harbor, 1998 is the same person) divided the study area into seven reaches based on tectonic position, slope, and channel pattern (Figure 6.12, Table 6.3). Reach 1 is designated as the control or unaffected reach, downstream of the anticline. Reach 7 is at the upstream margin of the uplift, and it is slightly entrenched below the older alluvial surfaces. Reach 6 crosses the upstream boundary of the axial graben and is the steepest reach in the study area. Reach 5 is located in the axial graben of the anticline, but the channel has been dramatically changed by recent bend cutoffs. Recently abandoned channels show that the former channel was meandering, narrower, and deeper. The present morphology probably reflects a response to cutoffs rather than uplift, and it may be short-lived. Reach 4 is located on the downstream graben shoulder, and it is less incised than reaches both upstream and downstream. Reach 3 is located at the downstream margin of the surface expression of the uplift, where the late Pleistocene surface becomes buried beneath the modern floodplain. Reach 2, which is arbitrarily split into two, is located where deposition is occurring below the zone of uplift.

The shape of the bankfull channel changes visibly through the zone of deformation. Thalweg depth varies irregularly through the uplift, but it remains roughly equal to that of the control reach (Figure 6.14A, Table 6.3).

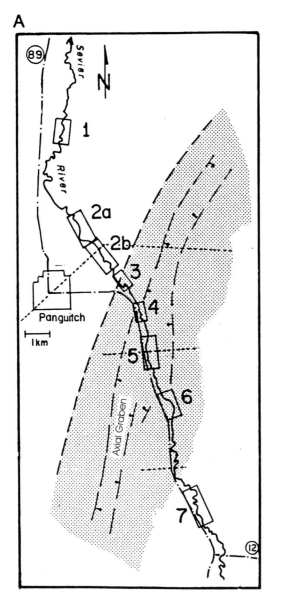

Figure 6.12 Map of Sevier River showing study reaches and location of cross-sections. Stippled area shows axis of late Quaternary uplift (from Jorgensen, 1990). Flow is to the north.

Aggradational areas (Reaches 2a and 2b, Figure 6.12) are less deep because of the wide, flat channel shape that is produced by lateral activity and the deposition of large bars. Reach 5 is shallow because of the recent channel cutoffs (Figure 6.13), which caused the river to incise and widen.

Channel width (Figure 6.14B) is greater and has greater variability in the

Table 6.3. *Quantitative summary of Sevier River response to deformation*

Reach		1	2a	2b	3	4	5	6	7
Sinuosity		1.59	1.34		1.21	1.12	1.10	1.26	1.49
Valley slope		0.0021	0.0043		0.0060	0.0040	0.0042	0.0061	0.0044
Bankfull channel characteristics									
Area	(m²)	29.5	31.0	22.5	21.9	22.0	20.1	29.5	26.5
Discharge	(m³/s)	26	38	28	47	45	36	56	31
Width–depth ratio		23.3	27.3	37.4	18.3	17.7	28.9	20.6	30.9
Width	(m)	36.8	40.7	44.2	26.7	26.6	32.0	31.8	41.5
Depth	(m)	1.58	1.49	1.18	1.46	1.50	1.11	1.54	1.35
Flow characteristics of the mean annual flood									
Water surface slope		0.0014	0.0028	0.0034	0.0043	0.0041	0.0053	0.0058	0.0023
Width	(m)	27.1	30.0	35.2	18.2	20.0	24.2	27.5	31.7
Thalweg depth	(m)	1.38	1.14	1.03	1.03	1.06	0.84	1.08	1.13
Average velocity	(m/s)	0.86	1.17	1.18	1.70	1.60	1.53	1.28	1.07
Boundary shear stress	(N/m²)	11.3	17.6	20.3	27.4	27.3	28.5	34.0	14.1
Unit stream power	(W/m²)	9.7	21.5	24.8	46.8	44.4	43.5	44.1	15.3
Sediment characteristics									
Median bar	(mm)	5.4	10.8	11.4	17.5	17.7	11.7	22.8	4.6
Median bed	(mm)	14.3	36.3	39.6	46.6	50.7	48.4	57.5	21.8
Excess shear-bed[1]	(N/m²)	8.6	4.4	6.1	9.9	6.7	6.1	20.6	9.2
Excess shear-bar[2]	(N/m²)	7.5	9.6	12.5	14.2	15.0	19.5	17.5	10.9
In-channel sediment	(m²)	39.4	32.1	54.6	26.2	22.4	46.8	33.2	28.7

Notes:
[1] Using maximum shear stress and D_{50} bed sediment size.
[2] Using average shear stress and D_{84} bar sediment size.

Downstream —————————————→ Upstream

Position	Reach 1	Reach 2	Reach 3	Reach 4	Reach 5	Reach 6	Reach 7
Position	control	below uplift	downstream margin of uplift	downstream edge of axial graben	in axial graben of uplift	upstream margin of uplift	above uplift axis
Vertical Change	stasis	aggradation	slight valley incision to stasis	deep valley incision	deep valley incision	deep valley incision	moderate valley incision
Channel Pattern							
Channel Change	slow meander extension and translation	rapid bank erosion, migration, and avulsion	slow meander growth and neck cutoff	slow lateral migration and meander growth	rapid bar growth, downstream and lateral bend development	u/s half is stable, but d/s half has had frequent avulsions	moderate bank erosion and chute cutoff
Bar Characteristics	narrow, low-relief, gravelly-sand, point and scroll bars	very wide, low-relief sand and gravel point and bank-attached bars, occasional large gravel bars in steep riffles	narrow, high-relief, well-sorted, gravel bars in a narrow slightly sinuous channel, large chaotic bars near cutoff	narrow, high-relief, well sorted, gravel bars in a narrow slightly sinuous channel, large coarse point bars in meanders	large low-relief bars that are mostly bank-attached, and narrow, less-exposed mid-channel bars	narrow, high-relief bars adjacent to pools and steep mid-channel bars in riffles	very wide, low relief point bars and long, narrow scroll or bank-attached bars, few chute bars
Pool-Riffle Sequence	deep scour holes at bend apices and shallow, flat riffles	deep scour holes at abrupt channel bends otherwise shallow, flat riffles and pools	very steep, cobbly riffles and deep narrow pools	steep, narrow riffles at heads of coarse-grained bars and deep, fast pools	deep pools in irregular bends wide long riffles with moderate sediment volume	deeply scoured pools with low water-surface slope and steep rapid riffles, irregular profile	deep, gentle pools adjacent to large, flat-topped bars, and flat-floored riffles
Bank/Sediment Characteristics	Occasional fan debris >200 mm but most sediment is fine-grained channel fill	low-relief banks with fine-grained cap over gravel base, easily erodible	mixture of coarse channel-fill and cobble-gravel bar deposits, left bank includes terrace deposits	large cobbles and boulders shed to the channel and floodplain from Sevier River Fm in adjacent slopes	mostly, low-relief, recently deposited bar material but occasional, resistant terrace and floodplain	very coarse cobble-gravel adjacent to channel and fine-grained terrace/valley flat	gravelly-sand in recent deposits and fine- to coarse-grained channel-fill in valley flat deposits
Channel Position	channel bank at elevation of valley-flat	bank profile oscillates below valley-flat profile	channel incised well below elevation of the valley flat	channel banks at elevation of valley flat, channel not incised	wide inner channel incised below valley flat, temporary incision	u/s half incised below terrace or valley flat, d/s half not incised	wide, incised channel beneath valley-flat remnants

Figure 6.13 Geomorphic characteristics of study reaches of the Sevier River (from Jorgensen, 1990).

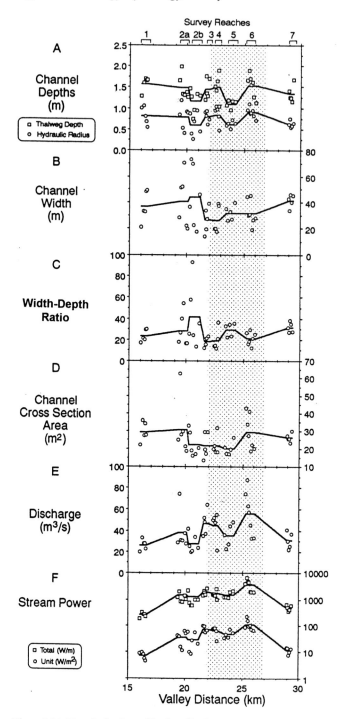

Figure 6.14 Morphologic and hydraulic characteristics of the Sevier River bankfull channel. The solid lines connect reach averages and stippling marks the uplift axis (from Jorgensen, 1990).

depositional reaches below the uplift than in the zone of incision. Change of width is the primary reason for a low width–depth ratio at the margins of the axial graben and a high width–depth ratio in Reach 2b (Figure 6.14C), where deposition and lateral activity are especially prominent.

Despite the long-term incision of the river, the bankfull channel area (Figure 6.14D) decreases in and downstream of the uplift, whereas the bank-full discharge increases (Figure 6.14E). The increase of bankfull discharge in a smaller channel reflects the steeper slope, and consequently higher velocity, on the capacity of the channel (Table 6.3).

The mean annual flood discharge (19.8 m³/s) was chosen as a representative, near-bankfull flow (Table 6.3). Channel velocity increases downstream through the axis of the uplift, reaching a peak in the narrow, deep cross-sections of Reaches 3 and 4 (Table 6.3). On the other hand, slope, as recorded by the water surface, decreases steadily through the entire uplift from the maximum in Reach 6. The channel bed in Reach 6 is more cobbly and rougher (higher Manning's n), and the channel is not as deep. These factors limit the increase of velocity despite the increase of slope.

Lastly, change of channel shape, water-surface slope, and velocity have a pronounced affect on the stream power available to transport sediment at the mean annual flood and at the bankfull discharge (Table 6.3, Figure 6.14F). Total stream power, is greatest in Reach 6 and decreases downstream except for a slight increase from Reach 4 to 3 (Table 6.3). Unit stream power also increases through the uplift but is a maximum in Reaches 3 and 4 at the downstream end of the incised reach (Figure 6.14B, C, D). Where unit stream power declines, deposition occurs.

The median size of bed and bar sediments decrease from Reaches 6 to 2, in comparison to the control reach and to the relatively low-gradient reach above the uplift axis (Table 6.3). The median bar grain size is coarsest where the channel incises the flanks of the uplift, with a slight decrease in the axial graben (Table 6.3).

In summary, the uplift produces a clear morphologic change of the Sevier River (Table 6.3, Figure 6.13), including a change of channel pattern from meandering to slightly meandering or straight, a change of slope with the steepest slope just upstream of the uplift axis, and an increase in flow velocity. Increased sediment transport or sediment supply increases the grain size of bed and bar sediment in the uplift (Table 6.3). Even the coarsest bed and bar sediment remains mobile throughout the study area. According to Jorgensen (1990), unraveling the interplay of these changes is difficult because the channel has responded in all of its components. The increased erosion and sediment transport, that mark the uplift, are related to a steepening and narrowing of the channel and an increase in the channel efficiency. Deposition is related to an increase in channel instability, profile irregularity (Figure 6.13), and channel widening.

Jefferson River, Montana

Like the Sevier River, the Jefferson River is a gravel-bed river flowing along the strike of a tectonic valley that has been deformed by more recent uplift. However, the Jefferson River is an order of magnitude larger than the Sevier River so the change of the stream morphology and behavior are correspondingly less. The Jefferson River profile is largely undisturbed through the uplift because changes of channel pattern, dimension, and shape produce the necessary difference in sediment transport to balance the effect of the uplift.

The Jefferson River lies in the Intermountain Seismic Belt of southwest Montana (Smith and Sbar, 1974), which is characterized by high seismicity, active extensional and strike-slip faulting, and surface deformation. Like most basin-and-range topography, the Jefferson River valley is tectonic in origin, formed by relative uplift and tilting of the Tobacco Root Range to the east and the Highland Mountain block to the west (Figure 6.15). Faulting and subsidence of the valley began in the Oligocene (Kuenzi, 1966), and faulting has continued to at least the latest Quaternary, although regional uplift has resulted in the general incision of valley fills (Reshkin, 1963). The Jefferson River in the study area encounters crystalline bedrock only at one location, where it flows adjacent to metamorphic rocks of the Highland Mountains just north of Hell's Creek (Figure 6.15). Elsewhere the river flows across the alluvial/lacustrine sequence of basin fill sediments.

The change of the Jefferson River in the area of tectonic deformation is apparent (Table 6.4, Figure 6.16) in the study area, which extends from below the confluence of the Big Hole, Beaverhead, and Ruby Rivers; to north of Silver Star (Figure 6.15). The Jefferson River is a gravel-bed stream with a meandering to a more irregular wandering channel pattern.

The Jefferson River crosses a late-Quaternary uplift at the center of an extensional valley. The location and degree of deformation are suggested by the warping of an alluvial surface, seismicity, and fault scarps. It is clear that the area of present uplift has been one of net erosion since at least the late Pleistocene, whereas aggradation or stasis predominate upstream and downstream of the uplift. The modern channel reflects this change as a decrease of sinuosity through the uplift, and as increased channel change and meandering in segments adjacent to the uplift. Jorgensen (1990) divided the study area into six reaches (Figure 6.17) based on channel morphology, channel

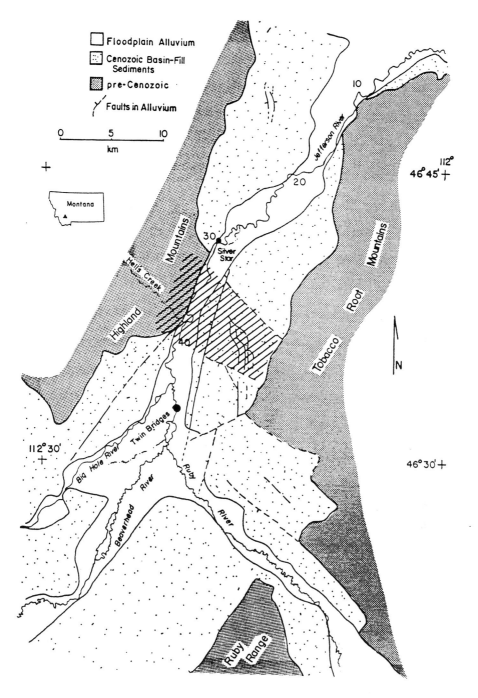

Figure 6.15 Geologic map of the Jefferson River study area. Geology from Ruppel and others (1983). Cross-hatched area is interpreted to be zone of maximum late Quaternary uplift. Faulting in alluvium north of Silver Star from Stickney and Bartholomew (1987). Flow is to the north (from Jorgensen, 1990).

Table 6.4. *Quantitative summary of Jefferson River response to deformation*

Reach		1	2	3	4	5	6
Sinuosity		1.83	1.38	1.18	1.10	1.13	1.30
Valley slope		0.0022	0.0026	0.0017	0.0019	0.0021	0.0020
Bankfull channel characteristics							
Area	(m²)	140	143	150	175	138	151
Discharge	(m³/s)	209	228	224	459	226	192
Width–depth ratio		35	39	35	32	39	46
Width	(m)	92	113	99	93	101	132
Depth	(m)	2.6	2.9	2.8	2.9	2.6	2.9
Flow characteristics of the 1.5-year flood							
Water surface slope		0.00121	0.00179	0.00149	0.00160	0.00189	0.00191
Width	(m)	617	213	335	76	440	512
Thalweg depth	(m)	2.63	2.81	2.80	2.26	2.59	2.83
Average velocity	(m/s)	1.49	1.65	1.49	1.70	1.54	1.37
Average shear stress	(N/m²)	17.9	20.6	20.0	20.4	26.4	19.1
Unit stream power	(W/m²)	26.6	34.3	29.9	35.2	40.1	25.9
Sediment characteristics							
Median bar	(mm)	16.3	21.2	27.8	19.4	17.4	20.4
Median bed	(mm)	53.3	60.7	70.0	64.9	75.0	56.5
Excess shear-bed[1]	(N/m²)	− 8.5	4.3	− 6.2	− 18.7	− 7.1	10.9
Excess shear-bar[2]	(N/m²)	− 0.8	− 2.4	− 9.7	− 0.5	5.4	− 4.4
In-channel sediment	(m²)	105	198	137	46	95	244

Notes:
[1] Using maximum shear stress and D_{50} bed sediment size
[2] Using average shear stress and D_{84} bar sediment size

planform, sediment storage, and channel activity or migration (Table 6.4, Figure 6.16). Reach 1 is the downstream control reach, Reaches 2 and 3 are below the uplift, Reaches 4 and 5 are in the uplift itself, and Reach 6 is above the uplift (Figure 6.17).

The surveyed channel profile of the Jefferson River is relatively smooth, and it decreases in slope downstream (Table 6.4). Clear changes of channel pattern in the study area are matched by changes in the characteristics of the bankfull cross-section (Figure 6.18, Table 6.4). Qualitatively, the channel in the uplift (Reaches 4, 5) is more symmetrical or regular, whereas the channel in aggradational reaches (Reaches 6, 3, 2) is characterized by irregular and large accumulations of sediment that often split the channel, and it is shallower on average than in the uplift and in Reach 1 (Figure 6.18A). However, almost no change takes place in the maximum or thalweg depth. The decrease of average depth or hydraulic radius (cross-sectional area/wetted perimeter) in depositional reaches is a result of the building of large bars, which increase

	Reach 1	Reach 2	Reach 3	Reach 4	Reach 5	Reach 6
Position	control	below uplift	below uplift	uplift	uplift	above uplift
Vertical Change	stasis	stasis or aggradation	stasis or aggradation	erosion	erosion	stasis or aggradation
Channel Pattern						
Channel Change	slow meander extension	rapid bank erosion, cutoff, and avulsion	downstream bar movement, slow lateral migration	very slow lateral migration and meander growth	slow bar growth and meander development	rapid bank erosion and channel avulsion
Bar Characteristics	large, coarse-grained point bars that fine downstream	chaotic accumulations of sediment in riffles and wide, high point bars	high volume, bank attached bars with less frequently flooded back-bar channels	low- to moderate-relief riffle bars with well-developed armor layer, very-low relief alternate bars	large bar-islands that are now mostly attached to the bank, narrow, minimally exposed point bars	large, high bars adjacent to pools that locally exceed bank height and steep mid-channel bars in riffles
Pool-Riffle Sequence	deep pools and rectangular, sediment-free riffles	narrow, deeply scoured pools and large, wide, sediment-rich riffles	deep, narrow pools and high, coarse, linear bars in riffles, very coarse bar crests	nearly indistinguishable pool-riffle topography, with constant channel width	deep cobble-armored pools and wide steep riffles with moderate sediment volume	deep gentle pools adjacent to large, flat-topped bars with riffles formed at connection of alternate bars

Figure 6.16 Geomorphic characteristics of study reaches of the Jefferson River (from Jorgensen, 1990).

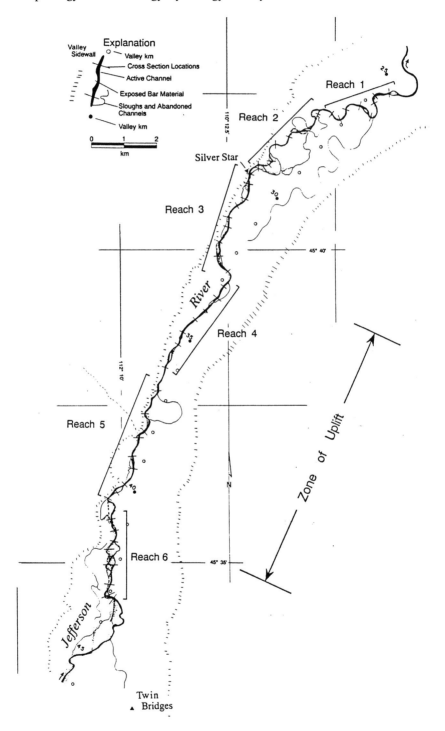

Figure 6.17 Map of the Jefferson River planform in the study area. Redrawn from U.S. Geological Survey orthophoto quadrangles (from Jorgensen, 1990).

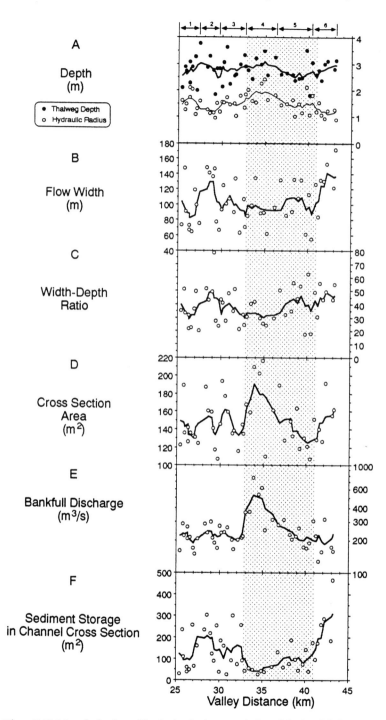

Figure 6.18 Morphologic and hydrologic characteristics of the bankfull channel. Lines are 5-point moving averages and stippling marks the zone of uplift (from Jorgensen, 1990).

channel width (Figure 6.18B) and channel asymmetry by forcing flow around the bar. Therefore, the bankfull width–depth ratio (Figure 6.18C), which is largest in Reaches 2 and 6 and least in Reaches 1, 4, and 5, reflects bar deposition taking place upstream and downstream of the uplift rather than incision at the axis of the uplift. The change of channel shape takes place without significant increase or decrease of the maximum dimensions of the channel. However, bankfull area and discharge (Figure 6.18D, E) are significantly larger in the downstream part of the uplift than either upstream or downstream. This difference appears to be caused largely by the minimal amount of sediment stored in the channel (Figure 6.18F) and channel deepening in the zone of uplift (Figure 6.18A).

The average bankfull flow for the study area is 255 m³/s, so a slightly lower flow with a 1.5-year recurrence interval (200 m³/s) was arbitrarily selected to represent the hydraulic characteristics of a single, near-bankfull flow (Table 6.4). Flow at this discharge is restricted to the channel at most cross-sections, but increased overbank flows typify reaches of deposition (Reaches 6, 3, 2), together with increased channel width and width–depth. Mean channel depth (Table 6.4) is low in Reaches 6 and 2 and highest in Reach 1.

With an equivalent or lower relative water surface slope, the channel becomes less rough in the straighter (Figure 6.16), sediment-free channel and produces a higher flow velocity (Table 6.4). Moreover, the lack of channel sediment increases the bankfull channel area and allows the channel to retain a greater portion of the discharge and thus stream power at increasingly large floods. These factors especially are responsible for transport of sediment out of the zone of uplift.

Above and below the uplift, where temporary and perhaps permanent deposition is occurring, the channel adopts a sinuous planform, and migrates extensively across the floodplain. This lateral activity distributes and reworks the oversupply of sediment. The change of channel pattern, and especially the change of bar type and position (Table 6.4), reflects not only the stability of the channel in terms of slope, discharge, and sediment transport, but also a metamorphosis of channel sediment size accompanying selective sorting and storage.

Discussion

A summary of Jorgensen's (1990) results is presented in Table 6.5. In order to simplify the presentation, Jorgensen grouped the responses into two types of river behavior, eroding and depositing. Erosion is associated with uplift or a steeper slope and deposition is associated with subsidence or gentler slopes.

Because the four rivers are subjected to different types of active tectonism and because each river is different, the only firm conclusion that can be reached is that deformation causes river variability. For example, under the category of valley and planform characteristics, there was no variable that shows a consistent change. Under the category of channel shape only width–depth ratio showed a consistent increase with deposition and a decrease with erosion. However, under the category of hydraulic variables, flow velocity, water surface slope, and stream power showed an increase with erosion and a decrease with deposition. As a result, under the category of sediment characteristics, sediment size increased with erosion, but sediment storage and bar size decreased with erosion.

The combined effects of channel shape and slope produce significant changes in unit stream power. Unit stream power is reduced in the subsiding reaches of the finer-grained streams (Neches, Humboldt) and increased in the reaches of greatest uplift in the gravel-bed streams (Table 6.5).

The most significant pattern change in all of the study areas is a change in the degree or style of meandering and bar building. Ouchi (1985) observed many of the same responses in experimental channels, where they were exaggerated because of the comparatively rapid deformation and recovery that occurred in the flume. The finer-grained rivers (Neches, Humboldt) increase sinuosity in reaches of steepened valley slope, as would be predicted from the established relationship between valley slope and sinuosity for laboratory channels (Figure 2.3). The gravel-bed rivers, on the other hand, are straighter in the axes of uplift and are most sinuous in depositional or control reaches, as suggested by Figure 3.3 for the steep braided channels.

The single most common sedimentologic response to tectonic deformation is the rapid fining of sediment in areas of deposition. Downstream fining takes place most rapidly on the Jefferson and Sevier Rivers, where coarse, well-sorted bars accumulate the material that is less easily transported downstream of the uplift.

The bar sediment of incising areas, including the finer-grained Neches River, is coarser as the competence of the stream increases with increasing slope, depth, or channel capacity. It appears that the bed of incising streams like the Jefferson and Sevier Rivers should become paved as the river cuts into the uplifted alluvium. A pavement (as distinct from an armor layer) is an immobile cover of coarse material that cannot be moved by the stream under most circumstances. Pavements frequently form below dams where sediment

Table 6.5. *Summary of the response of each study river to tectonic deformation (from Jorgensen, 1990)*

	Neches		Humboldt		Sevier		Jefferson	
	Eroding[1]	Depositing	*Eroding*	Depositing	Depositing	*Eroding*	*Eroding*	Depositing
Reach response	*Eroding[1]*	Depositing	*Eroding*	Depositing	Depositing	*Eroding*	*Eroding*	Depositing
Tectonic setting	*Uplift*	Subsidence	*Forward tilt*	Subsidence	Decreasing uplift	*Uplift*	*Uplift*	Subsidence or Stasis
Grain size	Suspended load (sand–clay)		Mixed load (gravel–sand)		Bed load (cobble–sand)		Bed load (cobble–sand)	
Response of valley and planform characteristics								
Planform	*Tight and irregular bends*	Two orders of meandering	*Tightly sinuous*	Broad, gentle meanders	Irregular bends and narrows	*Slightly sinuous*	*Straight to slightly sinuous*	Irregular, tight meanders
Sinuosity	*Increase*	Decrease	*Increase*	Decrease	Increase	*Decrease*	*Decrease*	Increase
Valley width	*Decrease*	Increase	*Increase*	Decrease	Increase	*Decrease*	*Decrease*	Increase
Valley slope	*Increase*	Decrease	*Increase*	Decrease	Decrease	*Increase*	*Decrease*	Increase
Migration rate	*Increase*	Decrease	*Decrease*	Increase	Increase	*Decrease*	*Decrease*	Increase
Response of channel shape								
Bankfull area	*Increase*	Decrease	*Decrease*	Increase	Decrease	*Increase*	*Increase*	Decrease
Bankfull width	*Increase*	Decrease	*Decrease*	Increase	Increase	*Decrease*	*Decrease*	nc[3]
Bankfull depth	*Increase*	Decrease	*Decrease*	Increase	Decrease	*Sl increase*	*Sl increase*	nc
Width–depth ratio	*Decrease*	Increase	*Decrease*	Increase	Increase	*Decrease*	*Decrease*	Increase

Channel shape	*Asymmetric channel with deep scours and prominent sandy, point bars*	Symmetrical but debris choked, muddy channel with abrupt bends	*Narrow, smooth shape and regular profile*	Wide, irregular channel and irregular profile	*Narrow, smooth with high-relief coarse-grained bars*	Wide, shallow channel adjacent to large bars, stepped profile	*Smooth, U-shaped channel with low pool-riffle relief profile*	Asymmetric, sediment-filled channel with high, bar, pooled-riffle relief
Response of hydraulic variables								
Flow velocity	*nc*	Decrease	*Increase*	nc	*Increase*	Decrease	*Increase*	Decrease
Water surface slope	*Increase*	Decrease	*Increase*	Decrease	*Increase*	Decrease	*nc*	nc
Bankfull discharge	*Increase*	Decrease	*nc*	Increase	*Increase*	Increase	*Increase*	Decrease
Stream power	*Increase*	Decrease	*Increase*	Decrease	*Increase*	Decrease	*Increase*	Decrease
Response of sediment characteristics								
Bed material	*Coarsens*	Fines	*[2]*	Fines	*Coarsens*	Fines	*Coarsens*	Fines
Bar material	*Coarsens*	Fines	*[2]*	Fines	*Coarsens*	Fines	*nc*	Coarsens
Sediment storage	*–[4]*	–	*Decrease*	Increase	*Decrease*	Increase	*Decrease*	Increase
Bar size	*–*	–	*Decrease*	Increase	*Decrease*	Increase	*Decrease*	Increase
Armoring	*–*	–	*Less well developed armor on bar surfaces*	Well-developed armor surface on large bars	*Uplift reaches are not armored in comparison to depositional reaches*	Uplift reaches are not armored in comparison to depositional reaches	*Uplift reaches armored in comparison to depositional reaches*	Uplift reaches armored in comparison to depositional reaches

Notes:
[1] The response of individual study reaches have been generalized to those that are eroding and those that are depositing over the long term. Eroding reaches are shown in italics.

[2] Result not clear.

[3] nc = no change is determined

[4] – = data not available

is eroded from the bed until only the coarsest, immobile grains remain. Bull and Knuepfer (1987) postulate that incision of the Charwell River of New Zealand following uplift resulted in the development of immobile, channel-bed pavements that were periodically disrupted by extreme events, which then renewed incision and created a series of terrace deposits paved with a coarse lag.

Another problem exists when the alluvial channel encounters bedrock at the axis of uplift. For example, on the Gulf Coast, bed and bank sediment samples were collected in the Bogue Homo channel in an effort to determine the effect of active tectonics on stream sediment properties. The results revealed that grain size distribution along Bogue Homo is controlled by the distribution of bedrock. The Hattiesburg clay, which locally forms the floor of the channel, is overlain by coarse gravels of the Citronelle Formation. The presence or absence of these gravels in the banks determines grain size distributions within the channels rather than hydraulic variability.

Figure 6.19 shows the D_{84} sediment size across the uplift for Bogue Homo. Initially, it appears that the abrupt coarsening of bed material at km 27 is a direct result of the uplift axis at km 27; however, the reach of coarse bed material is also the reach of bedrock exposure in banks of the channel. On Tallahala Creek, the relationship is similar, though less distinct; sharp changes in grain size are probably a reflection of the localized sources of gravel in the channel banks.

These examples imply that channel response to tectonic deformation is variable and that changes of channel morphology or slope need not correspond to tectonic boundaries. One must bear in mind the relatively small magnitude of deformation that is achieved by tectonics over the short term and look first to other causes of change. In addition, as demonstrated on the Gulf Coast, large rivers can adjust more rapidly to deformation, and this may confound any attempt to generalize the effect of active tectonics on rivers of different sizes as well as rivers of different types (Figure 2.4). We conclude that syntectonic river response is a reality and that it can be recognized by channel change, but obviously other factors can complicate the situation, and the investigator must be alert to this possibility.

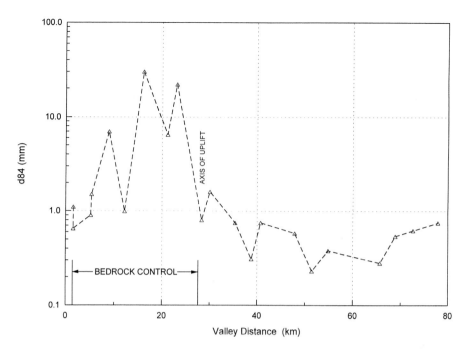

Figure 6.19 Sediment size (d84) variations in the bed of Bogue Homo.

7

_____

Lateral response

In the preceding chapters, the discussion has centered on uplift or subsidence, but lateral displacement is also of great importance for river response. Lateral shift of an alluvial river can be detected by its asymmetrical position in a valley and by evidence for repeated channel avulsion or shift in one direction (Figure 7.1). The channel may avulse suddenly down slope, or it may slowly migrate down tilt, or both cases can occur. In the first case, meander belts of decreasing age will be recognized down tilt. In the second case, oxbows of decreasing age will be identified on the alluvial surface. Braided streams are probably more likely to avulse than shift, as a result of high sediment loads and steep gradients. Examples are South Fork of the Madison River, Montana (Leeder and Alexander, 1987), Beatton River, British Columbia (Nanson, 1980b), and the Po River in northern Italy (Braga and Gervason, 1989) which show progressive shift, and the Carson River, Nevada (Peakall, 1998), Owens River California (Reid, 1992), and the Fly River, Papua, New Guinea (Blake and Ollier, 1970), which show avulsive change. Avulsive change is often related to episodic tectonic events (Chapter 5), although it is also a natural river response to aggradation.

An important concern for the structural geologist is, how reliable is a fluvial pattern for the identification of a specific structural pattern? In areas of low relief normal, reverse and strike-slip faults can give similar drainage patterns (Figure 7.2). The analysis of the drainage pattern is generally not enough to give evidence of fault movement and to quantify it, but consideration of the drainage pattern can provide information on the faulting activity.

Active strike-slip faulting

The most obvious lateral impact on a river is the result of active strike-slip faulting (Wallace, 1967). The geometry of a river channel crossing an

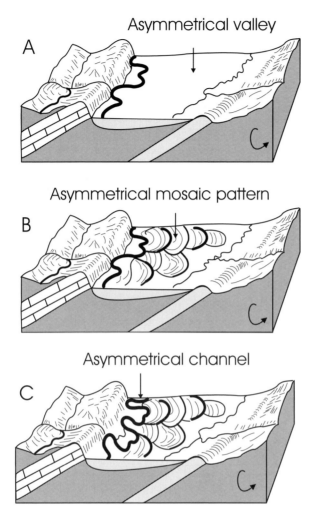

Figure 7.1 Diagrams showing different patterns of asymmetry: (A) the asymmetrical position of a river in a valley, (B) an asymmetrical meander belt, or mosaic of former channel positions, and (C) asymmetrical channel or meander development. Asymmetrical migration of a river may be continuous or avulsive. Continuous migration gives a pattern of directional scroll traces or oxbows with a concavity toward the migrating direction. Avulsive migrations are marked by successive abandonment of fluvial belts.

active strike-slip fault varies considerably according to location, local morphology, lithology, and of course, the movement of the fault itself (Figure 1.5).

Gaudemer *et al.* (1989) and Huang (1993) identified several types of channel pattern changes across a fault. Three types of river deflection are identified in Figure 7.3: with right-lateral river offset (R), left-lateral river offset (L), and unknown or no offset (U). Type 1 of classes R and L (Figure 7.3) reflect clear

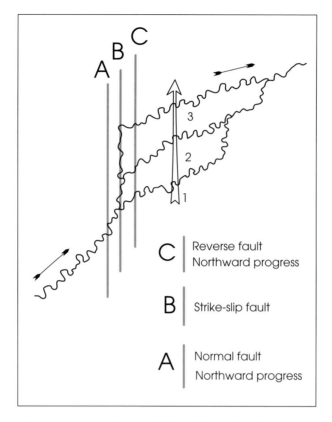

Figure 7.2 Pattern of lateral shifting of rivers, in relation to normal, strike-slip, or reverse faulting. The arrow shows the trend and the stages of directional shifts, from 1 to 3.

lateral offset. A variant of this type involves gradual deflection of the river before it intersects the fault. Types 2 and 3 of classes R and L show situations in which the geometry of the offset channels is more complex. Two values for the offset can be considered, a true offset (D) defined by the main trend of the channels to the fault, and the apparent offset (D'). The complex pattern is due to capture or deflection in the vicinity of the fault zone, perhaps due to the shattered material inside the fault zone. Rock crushed in the fault zone is more erodible, favoring this type of deflection. Thus, the true offset (D in Figure 7.3) is the most representative of the fault movement. Type 4 of classes R and L reflect cases where the deflection of the river near the fault is not in agreement with the observed offset. Deflections of a river opposite to the fault movement are reported by Gaudemer *et al.* (1989). Most of these anomalous cases reflect the capture of an upstream channel segment by another one located downstream of the fault, due to the fault movement. Class U are

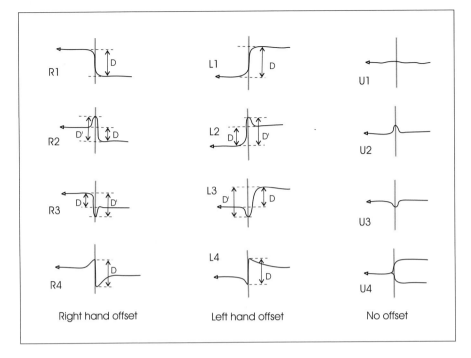

Figure 7.3 Classification of stream-channel offset, combined and modified from Huang (1993) and Gaudemer *et al.* (1989). R1 to R4: right-lateral deflected channels; L1 to L4: left-lateral deflected channels; and U1 to U4: undeflected channels. D is the true offset and D′ the apparent offset.

undeflected channels (Figure 7.3). Type U1 crosses the fault with no displacement. The channel may be recent and not yet offset by the fault, or it is the result of the junction of an upper channel with an existing or newly formed channel downstream. Huang suggests also that widening of the channel where it crosses the fault may hide all or part of the offset. The interpretation of U2 and U3 is complex, involving the combination of different river segments upstream and downstream of the fault, due to fault movement and capture. The type U4 involves the capture of one of two parallel streams.

Huang (1993) analyzed the active Yishi Fault in eastern China, and he concluded that half of the river offsets belong to the undetermined (U) class, and 75 percent of the remainder to the (R) class, suggesting a right-lateral strike-slip fault. Significant differences were observed from one segment of the fault to another with, for example, 90 percent right-lateral offset (R) and 10 percent unaffected (U) in one segment, and respectively 30 and 65 percent in another segment. Morphology, slope, and channel spacing seem to be responsible for most of these differences. Higher slope maintains straight channels, and close spacing aids capture. Gaudemer *et al.* (1989) obtained similar results from a

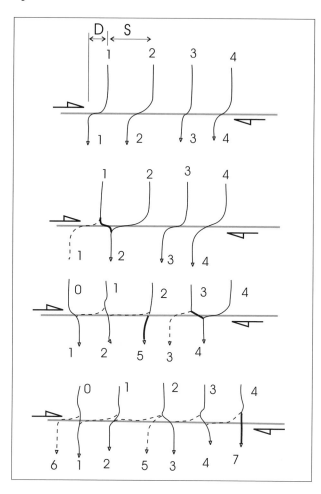

Figure 7.4 Model of successive capture and abandonment of a stream channel due to the offset of an active strike-slip fault (from top to bottom of figure). Dashed lines are the abandoned channel segments, and thick lines are newly developed segments. S is the interval between channels, and D the offset due to fault movement (modified from Huang, 1993).

study of the San Andreas fault, with only one-third of the rivers deflected in the direction of the fault movement. More than one-half of the rivers were not clearly offset by the fault, and 10 percent were offset in an opposite sense.

Stream capture generates patterns that may be difficult to interpret. For example, fault movement progresses with time, and successive channel capture and abandonment occur when an upstream segment joins an existing downstream drainage segment. Later these are separated, as the faulting progresses (Figure 7.4). The distance between channels has a significant effect on capture. Abandonment and capture occur easily when $D > S/2$, where D is the

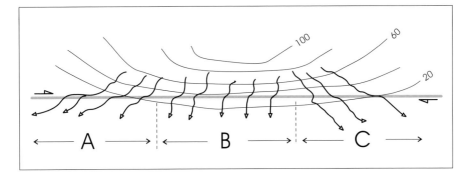

Figure 7.5 Effect of topography on the offset of the drainage network (from Huang, 1993). A: the effect of the morphology favors the offset of the fault; C: topography has a negative effect and can oppose the fault offset; B: no effect of topographic slope.

average channel offset and S the average interval between channels (Figure 7.4). Thus, the best place to measure fault displacement should be where both the slope and the channel spacing is small.

The degree of slope and the angle between slope direction and the trend of the fault make it either easier or more difficult for offset of rivers to develop (Figure 7.5). As mentioned before, a higher slope maintains straight channels, which increases the class of undeflected channels (U). The effect of the angle between slope and the fault is schematically presented in Figure 7.5. If the trend of the slope drives rivers in the same direction as fault offset, the resulting offset of rivers can correctly identify the fault movement (A area). However, if the slope trend is opposite with respect to fault deflection, a complex pattern is observed, which may incorrectly indicate fault movement (C area). Where the slope is steepest and the streams are normal to the fault, they should correctly indicate fault movement (B area).

As stated by Wallace (1967), the minimum channel displacements recorded may represent the displacement associated with the latest motion on the fault. Huang (1993) analyzed the adjustment of a river channel following faulting in piedmont areas where the river flows on alluvium. This is schematically represented in Figure 7.6. An initial straight stream, perpendicular to the fault is assumed (A). Fault movement offsets the channel, deflecting the channel along the fault. But the subsequent incision of the channel tends to shift the channel downslope off the fault, which opens an angle θ between the channel and the fault (B,C). If the fault stops moving for a long time, this angle increases as channel shifts down the slope (C, θ and θ'). However, when the fault moves again, the channel is offset once more, and another new deflection angle is formed (D). This process forms a segmented pattern of the channel

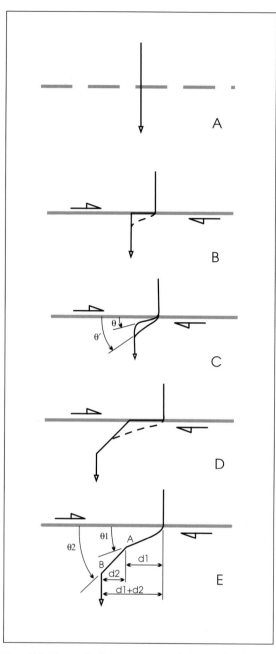

Figure 7.6 Channel adjustment on a piedmont border after offset by a strike-slip fault (from Huang, 1993). A: initial state, no offset; B: after offset, the channel tends to take the dashed line course, due to morphologic slope of the piedmont. During the following stages C to E the angles θ and θ′ represent the evolution of deflection through time. θ1 is the smaller and more recent deflection angle. d1 and d2 are the fault offset measured between the stream upper reach and the deflection points A and B. See text for details.

downstream from the fault, with angles θ1 and θ2 for the successive events. Thus, the deflection angle θ is closely related to the time of faulting. Fault motion produces a right angle at the fault, but stream incision and downslope shift between slip events changes it. Assuming that the angle of the stream with the fault increases linearly with time after faulting, Huang (1993) defines the following relation:

$$\theta = kT$$

where k is the incision rate and T is the elapsed time since faulting. In this relation k depends mainly on climate, lithology, and hydraulic conditions.

Wallace (1967) observed that large rivers are often offset more than small ones. This suggests that there could be a relation between river length and offset. If the rivers were antecedent (existing prior to the fault), large rivers and small ones should be theoretically offset by the same amount, equal to the cumulated displacement since the fault cut the drainage network. During an analysis of river offset along the San Andreas fault (Figure 7.7), Gaudemer *et al.* (1989) found a positive correlation between river length (L) and offset (δ) for 30 rivers crossing the San Andreas fault (Figure 7.8). This result appears to support the hypothesis that many of the streams are younger than the fault, and the youngest and shortest streams show the least displacement.

Half and uneven grabens

The asymmetric position of a river in a half graben or a graben may be the first indication of tectonic influences. This does not mean that the deformation is still ongoing, but that tectonic control of the river is probable. Russell (1954) interpreted the asymmetric position of the Menderes River in the Menderes graben in the Aydin area of Turkey as an indication of the active tilt of the graben surface, which appears consistent with more recent data on the neotectonic evolution of this region (Angelier *et al.*, 1981). Asymmetrical positioning of a river in a valley cannot always be related to active lateral tilt of the valley surface. For example, a river can be maintained along one side of a valley floor by sedimentary input from tributaries on the opposite side. The Basin and Range of western United States is an exceptional case of linear mountains separated by broad structural valleys (Pardee, 1950; Hunt, 1979). The asymmetrical position of rivers in most valleys of the Basin and Range is related to half grabens formed by tilted blocks along normal listric faults (see Stewart, 1978, for a synthesis). As early as 1912, Lawson observed that the valley floor is depressed near active faults. Lakes and rivers hug the fault. Where the faulted border of the valley is less active, the sediment delivered

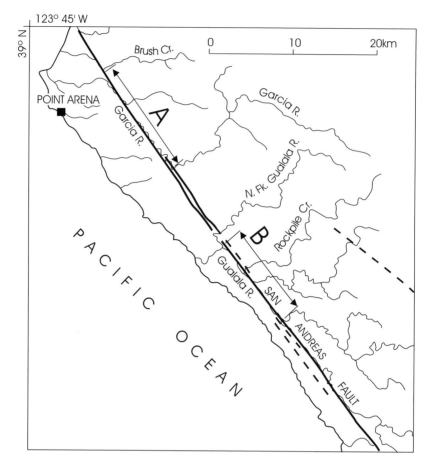

Figure 7.7 Offset of main rivers by the San Andreas Fault (Modified from Gaudemer *et al.*, 1989). A and B show examples of the maximum offsets that are representative of fault movement.

from the mountain front shifts the stream away from the mountain front, and the lowest part of the valley is located toward the center of the valley (Lawson, 1912, p. 95).

The August 17, 1959 earthquake of magnitude 7 in the Hebgen Lake area of western Montana is the strongest historical event registered in the Basin and Range (Doser, 1985). A comparison with precise leveling made before the earthquake shows evidence of tectonic tilting (Myers and Hamilton, 1964). Some modifications of the drainage network are also reported. Leeder and Alexander (1987) studied the floodplain of the Madison and South Fork Rivers, in the central part of the Hebgen Lake fault block. Both rivers are of medium size, with channel widths ranging from 5 to 50 m, and meander wavelengths are not over 250 m. The floodplain is not directly bounded by

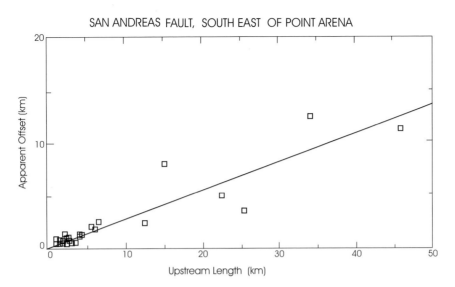

Figure 7.8 Relation between apparent offset and channel length of the rivers measured upstream from the San Andreas Fault southeast of Point Arena (from Gaudemer *et al.*, 1989).

faults. The progressive and directional migration of the Madison and South Fork Rivers occur on a surface formed by glacial outwash deposits. The flood-plain of the Madison and South Fork Rivers shows traces of meander loops concave toward the present position of the river (Figure 7.1B). Leeder and Alexander (1987) consider this pattern as a critical evidence of progressive directional displacement of the river channels (Figure 7.9). The total amount of river migration is 1600 m, starting about 7000 BP (Nash, 1984), which yields a mean displacement of 23 cm/yr. The results of Leeder and Alexander's study (1987) emphasize a steady directional river migration in the direction of tilt, which supports a structural interpretation of river shift. As another example, the Nile River hugs the eastern side of its valley, which suggests that the Nile valley is a half graben (see Chapter 10).

River shift in tectonic basins

The great alluvial plains are often characterized by a dense network of paleochannels, representing former positions of the present rivers. Whether the shift from one position to another was random, or controlled by structural trends and active tectonics, can be disputed.

The basins, where avulsion has been studied, are characterized by decrease of the channel slope at their entry, and deposition of coarse sediments as a result. This causes vertical aggradation of the channel, and subsequently its

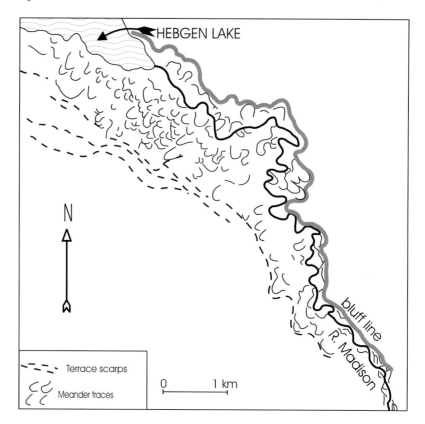

Figure 7.9 Asymmetric meander belt and lateral directional shift of the Madison River, south of the Hebgen Lake (modified from Leeder and Alexander, 1987). See details in the text.

abandonment for a lower position in the floodplain. However, river shifts in foreland basins are different, especially regarding the scale and the trends of the process. Channels shift up to several tens of kilometers, far beyond the limits of the active meander belt. In all documented cases, the phenomenon is repetitive and directional during the Holocene, which produces parallel paleomeander belts.

The analysis of river shifts is made with the combination of subsurface structural data and neotectonic and seismotectonic data from the basin and its margins. Most of the available cases of river shifts in flexural basins are characterized by the migration of a deflection point of the river along structural trends.

Subandean basins

The Marañón and Beni Basins in Peru are asymmetric troughs that formed as a result of the eastward motion of the Andean belt with respect to

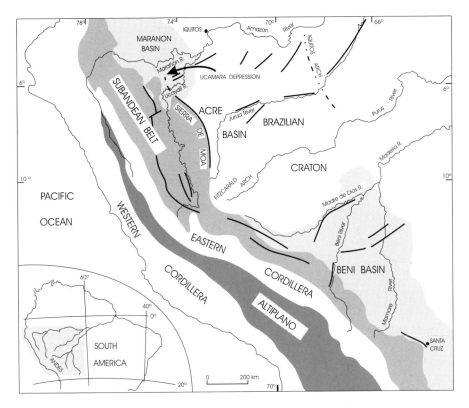

Figure 7.10 Location map of the Marañón and Beni foreland Basins.

the Brazilian craton (Figure 7.10). The loading of the edge of the continental lithosphere by the overriding orogenic belt produced subsidence (Beaumont, 1981; Jordan, 1981). In western South America, earthquake focal mechanisms show approximately E–W (N80°E) compressional stresses, which are observed both in the Subandean region and in areas of low altitude (Assumpçào and Suarez, 1988; Assumpçào, 1992).

The Marañón Basin extends over about 400 km west to east and 500 km northwest to southeast. It has two parts: the Pastaza Depression to the northwest (Laruent and Pardo, 1975), and the triangle-shaped Ucamara Depression (Villarejo, 1988) to the southeast and east. The Ucamara Depression is a vast floodplain of about 25 000 km², where three main rivers join near the outlet of the basin, forming the Amazon River (Solimoes in Brazil). They are the Marañón and the Ucayali Rivers draining from the Andean range, and the small Tapiche River (Figure 7.11).

A dense network of small and large streams drains the Ucamara Depression. Precipitation inside the depression generates "black-water" streams, rich in organic acids, whereas the rivers coming from the Andes have

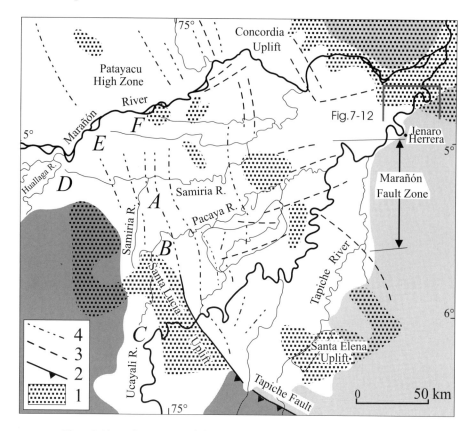

Figure 7.11 Drainage map of the Marañón Basin superimposed on basement structures (from Laurent, 1985 and Laurent and Pardo, 1975). (1) Paleozoic uplifts, (2) Tapiche Fault, which is the main front of the Andean overthrust tectonics, (3) major pre-Tertiary faults in the basement, suspected of reactivation by the Andean tectonics, (4) axes of folds related to Andean tectonics. Dark gray is the foothills area, and light gray is the Brazilian Craton margin. A, B, and C show the successive stages of migration of the deflection point of the Ucayali River, and D, E, and F are the correlative stage of migration of the Marañón River (from Dumont, 1996).

"white" silty water. Several small black-water streams cross the depression, from the foothills to the craton margin. The longest are the Pacaya and Samiria Rivers (Figure 7.11), which are old courses of the Ucayali and Marañón Rivers. A lower terrace with small channels was dated at 13 000 BP. As a result, all the channels attributed to recent migration of the Marañón and Ucayali Rivers and related tectonics are probably more recent than 13 000 BP. Remnants of these oldest channels are preserved in some remote places of the Ucamara Depression.

As deduced from the position of the underfit streams, the Marañón River shifted 50 km to the north, and the Ucayali River shifted more than 100 km to the south and southeast of a previous common trunk (Figure 7.11). The process involved successive directional avulsions, each of 10 to 30 km. Entering the depression, the Ucayali River is deflected to the east. Previous paths of the river show a southward upstream shifting of the deflection point (points A, B, and C, Figure 7.11), while 30 to 50 km avulsions occurred in the central part of the depression. There is no indication of increasing slope: the sinuosity of the river channel does not change upstream and downstream of the deflection point, and no sedimentary fan system forms in relation to the deflection. In the central part of the depression, the drainage network trends east–northeast, with some short left-lateral deflections to the north linking the main segments (Figure 7.11). Contemporaneously, the Marañón River was deflected to the north to occupy the lower reach of the Huallaga River (D, E, F, Figure 7.11). The straight course of the former Huallaga River suggests fault control.

The structural style of deformation at the front of the Peruvian Subandes involves flat overthrusts with frontal listric upthrusts (Pardo, 1982; Megard, 1984). A model of outward progress of foreland structures and the effect on depocenters (Zoetemeijer *et al.*, 1993) shows that piggy-back basins developed behind the tectonic front can control the rivers issuing from the foothills, by keeping them flowing along the axes of piggy-back depocenters. According to this model, the Tapiche Fault controls the southward shift of the deflection point of the Ucayali River (positions A, B, and C, Figure 7.11).

In the central part of the Ucamara Depression, the orientation of the small (underfit) and present (large) rivers are parallel to the main trend of the identified faults of the Marañón Structural Zone (Figure 7.11). A similar direction is observed for the main set of normal faults reported from the Craton margin. This suggests that active tectonics probably controls the main trend of the drainage network in the central part of the depression.

Lower Ucayali River in Peru

The flat surface of the Marañón flexural basin extends over 400 km west to east and 500 km north to south. The basin is topographically delimited toward the Brazilian Craton by uplands. Despite low relief (about 20 m over the basin surface), the drainage of the basin has only one exit, toward Iquitos. The basin surface near the exit is a wide floodplain common to the Ucayali and Marañón Rivers, and it is called the Ucamara Depression (Figure 7.10). More precisely, the two rivers join on the eastern margin of the basin in a triangle-shaped area north of Jenaro Herrera (Figure 7.11) called locally the Nauta Depression. Cohesive silt and clay of the Pebas Formation of late

Pliocene to early Quaternary age underlie the alluvial sediment of the present floodplain (Dumont *et al.*, 1991). The combination of fault and fluvial pattern analysis indicates that the Nauta Depression is a graben. The analysis of the asymmetrical pattern of the Ucayali River at several scales, including the more recent development of the meanders, suggest that deformation is still ongoing.

A map of the floodplain allows one to assign a rank to each element of the channel patterns, according to their relations with the surrounding elements. For example, a particular element of the pattern is ranked younger than the element it cut, and older than the element cutting it. The main pattern suggests lateral migration of the channel of the Ucayali River 5 to 15 km toward the upland border. Successive stages of meander construction are illustrated by the mosaic pattern (Figure 7.12). The ages obtained for the five major stages of mosaic construction are: (1) the present active meanders, (2) 140–190 BP, (3) 340–520 BP, (4) 990–1310 BP, and (5) 2100–2310 BP. These data suggest episodic changes of the channel.

A study of the channel using 1960 aerial photographs, 1970 Landsat, and 1988 Spot images suggests a 60 m/year migration rate of the apex of a meander. A meander loop of 7 km to 10 km forms in between 100 years to 160 years according to this rate. Using testimony from local settlers, Lamotte (1990) gives an estimate of 25 m/year of lateral accretion during 40 years for a part of meander located between the apex and the inflection point, in the same area. Hence, one may estimate 100 years to 200 years for the formation of a meander loop. This estimate is based on an assumption of a similar rate in the past and present. As a result, the five stages of mosaic construction occupy only ¼ to ⅓ of the total period of time available.

According to paleoclimatical records from western and southwestern Amazon, the construction of the fluvial mosaics occurred during a wet climate. During dry periods, floods in the channel of the Tapiche river did not exceed the present low water level (Dumont *et al.*, 1992). During periods of lower discharge the channel was probably underfit, more or less fixed or migrating more slowly. The onset of a wet period allowed the channel to recover its independence, migrating toward the more depressed area. Thus, dry periods appear to block migration and wet periods to favor it, and migration events, therefore, cannot always be related to tectonics.

Beni Basin

The Beni Basin extends over about 800 km of latitude, broader to the northwest (500 km) than to the southeast (150 km), (Figure 7.13). It is located on the southern margin of the Amazon rain forest. Three large meandering

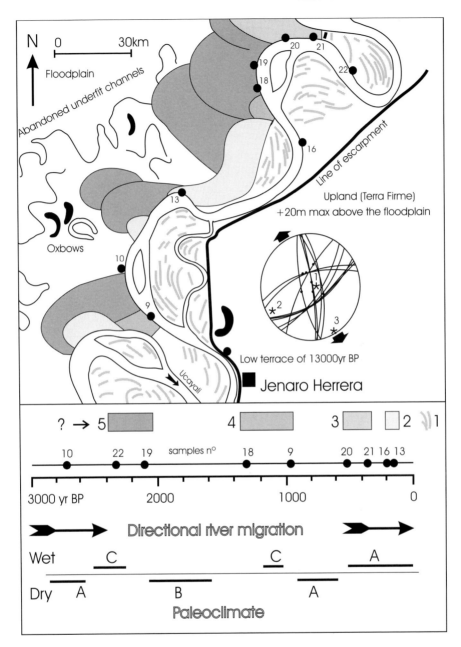

Figure 7.12 Lateral directional shift of the Ucayali River near Jenaro Herrera (Figure 7.11) as evidenced by the age of components of the mosaic pattern. Age of the mosaic elements are 1: ridges and swales of the present stage of mosaic construction; 2: 140–190 years BP; 3: 340–520 years BP; 4: 990–1310 years BP; 5: 2100–2310 years BP. Numbers on the map are points of collection of samples, age in years BP: (9): 900 ± 40 years BP; (18): 1310 ± 40 years BP; (19): 2100 ± 40 years BP; (20): 520 ± 50 years BP; (21): 340 ± 50 years BP; (22): 2320 ± 40 years BP.

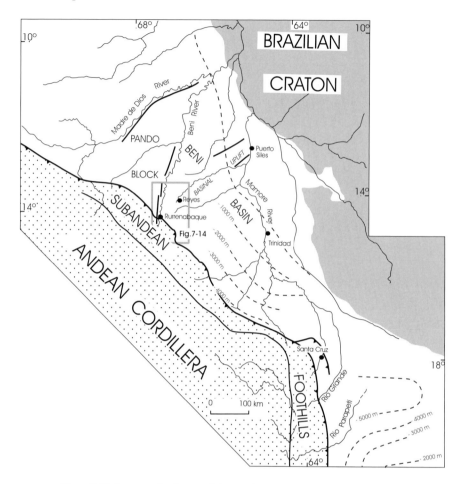

Figure 7.13 Structural scheme of the Beni Basin. Note the former northeast courses of the Beni River between Rurrenabaque and Puerto Siles, now occupied by a basinal uplift. Detail of the lateral shift of the Beni River is shown in Figure 7.14. Dark rectangle shows location of Figure 7.14.

rivers cross the basin; from north to south they are: the Madre de Dios, the Beni, and the Mamore Rivers. They join in the northeast of the basin, to form the Maderia River, which exits the basin on gneiss outcrops. The most depressed area, where most of the wetlands and fluvial traces are reported, is between the Beni and Mamore Rivers.

Plafker (1964) identified a counterclockwise shift of the Beni River, and Allenby (1988) interpreted the phenomenon in relation to a basinal uplift trending northeast from Reyes to Puerto Siles (Figure 7.13). Crossing the uplift area, the channel of the Mamore River shows a significant reduction of sinuosity, and it is interrupted by several rapids north of Puerto Siles. The axis of the structure is parallel to the main trend of lineaments in the Brazilian

Craton (Plafker, 1964; Allenby, 1988). This lineament extends in Central Brazil toward the Tapajos neotectonic fault (CERESIS, 1985).

Dumont and Hannagarth (1993) identified five successive stages of shifting of the Beni River downstream from Rurrenabaque. The oldest traces (A and B, Figure 7.14), follow partly the present position of the Yakuma and Tapado Rivers, which join the Mamore River. The channel traces of these stages, and principally those of the Tapado stage, are smaller than the traces of the Beni River. The younger stages (C, D, and E, Figure 7.14) follow the Yata, Biata, and Negro Rivers, which are continuous streams with channel width and meander wavelength similar in size to the Beni River.

The successive shifts of the Beni River defined a northward downstream migration of the deflection point of the river entering the basin. The early stages of the Beni River turned abruptly to the northeast at the outlet from the foothills, downstream from Rurrenabaque. The point of deflection shifted downstream inside the basin (C′, D′, and E′, Figure 7.14). This shows that the shifting of the Beni River was initiated at the foothill border, and it progressed downstream by successive increments.

The pattern of fluvial shifting, which involved the migration of the inflection point, suggests faulting. Nevertheless, the rainforest in the foothills as well as the wetlands near the river make the identification of the fault difficult. Direct evidence comes from north-striking faults in an upper terrace of the Beni River, with an offset of some decimeters, and an upthrow of the western side. Also, morphological differences are observed near the piedmont, on both sides of the outlet of the Beni River from the foothills. Northwest of the outlet, the small streams carrying pebbles flow along the slope in incised channels (Figure 7.14, area 1). Southeast of the Beni River the landscape is very different. The streams sloping down from the piedmont wander on the surface of very flat and sandy alluvial cones; the floodplain of the Beni River is very close to the piedmont (Figure 7.14, location 2). This morphology suggests that the area south of the outlet of the Beni River is depressed with respect to the area to the north.

There is an indication that the Beni River crosses an area of accelerated subsidence about 50 km north of Rurrenabaque. The floodplain rises just 2 m above the low water level. These are the lowest river banks observed along the Beni River in the basin. Accelerated sedimentation buries large areas of drowned forest. The transit of gold particles, which are found upstream from Rurrenabaque, almost stops in this area (Dumont *et al.*, 1991b). Also, the Buenaventura fault is consistent with the observation of active subsidence along the margin where there are lakes.

Morphological and neotectonic data suggest that the Beni River was forced

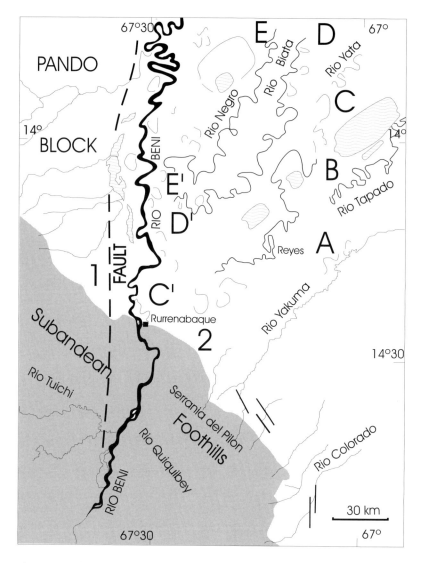

Figure 7.14 Migration of the Beni River north of Rurrenabaque (Figure 7.13). A to
E are the successive positions of the Beni River (the oldest position A is rather
speculative). C′, D′, and E′ are the successive positions of the deflection of the
Beni River, corresponding to stages C′, D′, and E′ in the basin. Faults are observed
in the foothills and inferred in the basin. Areas 1 and 2 are respectively uplifting
and subsiding, relative to the floodplain surface (from Dumont, 1996).

to the north by a north-striking fault which forms the border between the
uplands of the Pando Block to the west and the Beni River floodplain to the
east. The process involves a more active subsidence to the east, channelizing
the Beni River along the fault.

Pannonian Basin, Hungary

Mike (1975) studied the relationships between the different stages of paleochannel traces in the eastern Pannonian Basin. He interpreted these traces as successive positions occupied by the Tisza River. He also interpreted the successive shifts of the river as controlled by active tectonics, but the basement structures and neotectonics of the Pannonian Basin were too poorly known to support Mike's interpretation. Mike's fluvial data were reinterpreted using recently published structural and neotectonic data. Fluvial and geomorphologic data come from Mike (1975). Information on subsurface structures are from Royden (1988) and Royden and Horváth (1988), seismotectonic data are from Rumpler and Horváth (1988) and Müller *et al.* (1992), neotectonic data are from Horváth (1988, 1993) and Bergerat (1988), and recent vertical surface movements are from Joó (1992).

The Pannonian Basin is a Cenozoic back-arc basin located inside the loop formed by the Carpathian Belt (Figure 7.15). The basin comprises a complex patchwork of transtensional pull-apart and rift basins (Royden and Báldi, 1988). Pliocene and Quaternary deposits overlay and unify these small tectonic basins. The Danube River (west) and the Tisza River (center) drain the central part of the Pannonian Basin, also called the Great Hungarian Plain. The middle Tisza River lies over the Szolnok flysch basin, a Paleogene basin compressively deformed with imbricate thrust and faults (Nagymarosy and Báldi-Beke, 1993). During middle Miocene times (17.5–10.5 Ma) molasse deposits filled half grabens and pull-apart basins reflecting syn-rift structures. From late Miocene times to Present (10.5–0 Ma) a post-rift sedimentary succession covered the previous deposits, constituting the uppermost sedimentary unit. According to Royden and Horváth (1988) this widespread sedimentary cover results from regional thermal subsidence. A belt of high density of faults, the mid-Hungarian Fault Zone (Csontos *et al.*, 1992), crosses the Pannonian Basin from southwest to northeast (Figure 7.15). This is a left-lateral wrench fault zone initiated during the end of early Miocene time (Tari *et al.*, 1993), and it was reactivated as left-lateral strike-slip faults during late Miocene time (Sztanó and Tari, 1993).

In the area located between Debrecen to the northeast and Szeged to the southwest, Mike (1975) identified eight successive sets of fluvial traces attributed to the paleo courses of the Tisza River (a–h, Figure 7.16). All these traces have similar size except one, the set "g" with smaller traces, which are correlated with the cold boreal period between 9000 and 7500 BP (Mike, 1975).

It is possible to group the eight sets of fluvial traces into three main stages

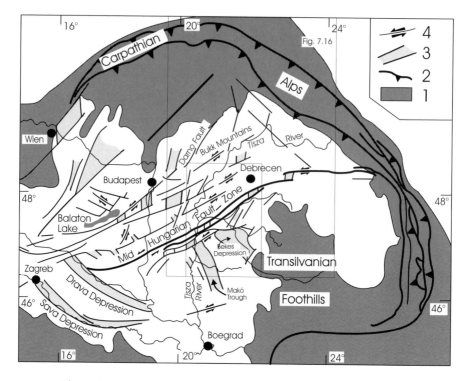

Figure 7.15 Structural scheme of the Pannonian Basin, modified after Rumpler and Horvath (1988), simplified for the Carpathian Belt. 1: Alpine–Carpathian (north) and Dinaride (SE) mountains. 2: thrust faults; 3: areas of major crustal extension and subsidence; 4: strike-slip faults.

(Figure 7.16). Stage I includes the sets (a), (b), and (c). Stage II includes the three sets (d), (e), and (f) near and south of Debrecen, and Stage III includes the recent sets (g) and (h), near the present position of the Tisza River.

During Stage I, the Tisza River was flowing from north to south near the border of the Transylvanian foothills, then it made a westward deflection over the Bèkès Depression. The deflection of the river toward this Cenozoic graben suggests that extension was effective. The following Stages II and III are the result of two important directional avulsions to the northwest. During Stage II, river traces are parallel to and superposed over the mid-Hungarian Fault Zone. The avulsion to Stage II is interpreted as a result of the renewal of compressive stress during the Holocene, and the left-lateral reactivation of the mid-Hungarian Fault Zone, that may have formed elongated structural ridges and depressions that were able to capture the river channel. The avulsion to Stage III originated far upstream near the Bükk Mountains piedmont and occurred during Boreal times. According to these data, tectonic oscillations

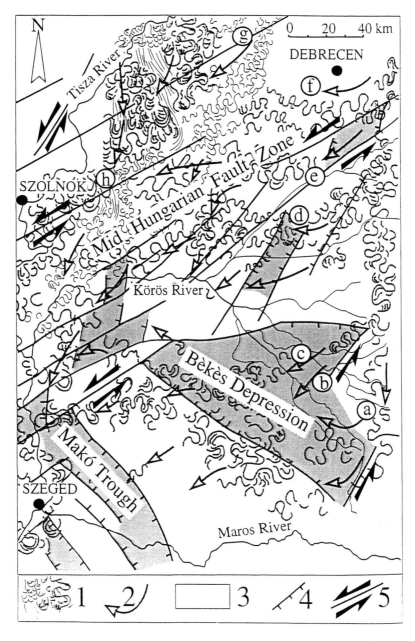

Figure 7.16 Fluvial traces of the Tisza River (modified after Mike, 1975). The letters (a) to (h) refer to the successive sets of traces from the oldest to the youngest. Arrows show local trend of paleodrainage. Shaded areas are major crustal extension and subsidence areas of Miocene. Dark arrows indicate direction of movement of strike-slip faults. See location on Figure 7.15.

from extension to compressive stress are responsible for the shift of the Tisza River in the Pannonian Basin.

Indus River avulsions

The Indus River in Sindh Pakistan exhibits a range of modern channel patterns, and the wide alluvial plain displays numerous ancient courses of the river (Figure 7.17). The river changes from braided to anastomosing to meandering as it crosses several apparently active structures (Figure 7.18). Obviously, the river has avulsed across the alluvial plain many times in the relatively recent geologic past. Of particular interest is the fact that there appear to be two locations where avulsion has occurred repeatedly, one is upstream of Kandhkot and the other downstream to the east of Manchar Lake. Both of these locations are near major tectonic boundaries (Jorgensen *et al.* 1993; Harbor *et al.*, 1994; Flam, 1993), and it is reasonable to conclude that avulsion in this case is not the result of cross-valley slope variations, but it occurs at specific sites that have tectonic controls. As a result of these controls the modern Indus plain can be thought of as two inland alluvial fans, one focused north of Sukkur and the other near Sehwan. Avulsion occurs at the apex of these fans. Indus River behavior suggests that the location of avulsions may be controlled by tectonics. This conclusion is reasonable because avulsions occur repeatedly at the same location.

Discussion

The deflection of a river course in successive directional avulsions along a fault is the most characteristic pattern observed in subsiding basins. However, the fault along which the river is deflected may not be active, because topographic measurements and records of seismotectonics actually are generally lacking. Nevertheless, the faults involved in the process of redirecting the fluvial drainage in the Subandean Basins were all active during the Quaternary, and they are compatible with the present seismotectonic activity of the regions under consideration.

In the Subandean Basins, decrease or renewal of stress from plate convergence probably controls the distal or proximal position of the confluence between the Marañón and Ucayali Rivers. A similar analysis for the Pannonian Basin indicates that the change is from extensional to compressive tectonics. The Tisza River changed from being trapped by depressions that were controlled by extensive tectonics to control by transcurrent lineaments. The ultimate shift of the river was controlled by a fault on the piedmont margin (Bükk

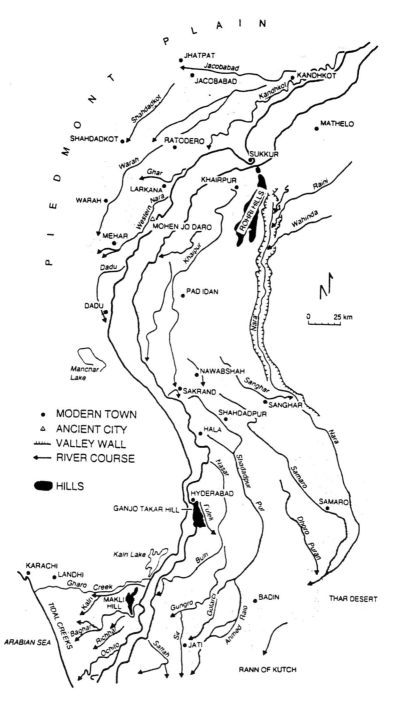

Figure 7.17 Present and ancestral courses of the Indus River in the Holocene (modified after Flam, 1993).

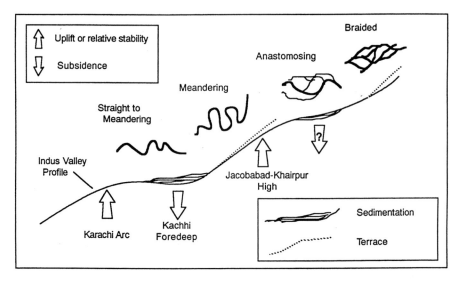

Figure 7.18 Cartoon showing the relationship of the Indus valley profile and channel pattern to tectonic elements. Note that the slope and pattern changes are exaggerated for illustration (from Jorgensen *et al.*, 1993).

Mountains border) illustrating the increase of Carpathian tectonics. The Indus River avulsion sites appear to be controlled by major subsurface tectonic elements of the Indus alluvial plain.

Climate also has a direct effect on the water discharge of a river, and subsequently on the size and shape of the fluvial traces. Increase of precipitation generates higher discharge, forming longer wavelengths, and larger channel width (Dury, 1970; Schumm 1977, 1986; Baker, 1978), but a dry climate limits river discharge and mobility. Thus, it has a negative effect on river dynamics in areas submitted to surface deformation. Wohl and Georgiadi (1994) correlated periods of meander abandonment of the Sejm River in the Dnieper Basin with episodes of increased flooding. In the floodplain of the lower reach of the Ucayali River, migration during the last 3000 years was active principally during periods of wet climate. A short period of dry climate will have little effect on river mobility because of a reduced discharge. Higher mobility occurs at the onset of warm and wet periods. In tectonic basins the increase of river mobility due to the onset of a wet period will drive rivers toward topographic lows or areas of active subsidence. As a consequence, major river avulsion will probably occur during, or at the onset, of a wet climatic period (Wohl and Georgiadi, 1994), although the cause is tectonic.

Part IV

Applications

Tectonics and fluvial sedimentation

In Chapters 2, 3, and 4, the response of a stream pattern to changing valley slope was stressed as an indicator of active tectonics. Channels can change from straight, meandering, braided, and anastomosing to any other pattern. Obviously, fluvial deposits will change as the pattern changes. Although channels may adjust to slope by changing pattern, they may also adjust by avulsion to another location on the floodplain. Also, river behavior can change from relative stability to aggradation, degradation, lateral shift, or avulsion. All of these responses cause a change in sediment transport and the location and type of deposition. This chapter is a discussion of how fluvial deposition is controlled by active tectonics.

Tectonically forced changes of fluvial deposition can be reconstructed by facies analysis of the resultant strata. For example, a decrease of slope during episodes of subsidence was cited as the cause for an upward shift from steeper braided to lower-slope meandering streams in the Triassic Chinle Formation of the U.S. Colorado Plateau by Blakey and Gubatosa (1984). In contrast, subsidence of the basin floor without rapid aggradation might increase relief, promoting preservation of steeper braided deposits during periods of high subsidence (García-Gil, 1993). Increased relief between basin and source areas may be sufficient in some cases to alter patterns of basinal fluvial systems by changing sediment supply. For instance, oversupply of coarse sediment owing to relative uplift in flanking thrust sheets in the retroarc Bowen foreland basin, Queensland, Australia caused an upward transition from low sinuosity meandering streams to braided streams in the late Permian Rangal coal measures (Fielding *et al.*, 1993). Numerous lithofacies models and architectural-element associations have been developed to help ascertain parent river characteristics from fluvial deposits. A review of these studies is far beyond the scope of this chapter; however, they have been thoroughly considered by Miall (1996).

Attributing vertical or lateral variations of stream characteristics within a fluvial deposit to variations of slope due to tectonics is difficult, and it requires elimination of other factors. For instance, channel pattern and size variations can be due to changes in discharge and sediment load instead of slope (Schumm, 1968; Ethridge and Schumm, 1978). Likewise, comparisons of flora and fauna among deposits could provide indications of climate change (e.g., Blakey and Gubitosa, 1984). Attribution of pattern changes to uplift or subsidence must be handled on a case-by-case basis.

There are two primary ways in which active tectonics can influence rivers, and hence fluvial deposits. The first is by a change of slope (a decrease, an increase, or lateral tilting). These changes can lead to pattern adjustments and/or avulsion. The second change is when an uplift produces increased quantities of coarse sediment that influences river morphology and behavior at some distance from the site of deformation.

Effects of tilting

Tilting may be considered in terms of either lateral (normal to floodplain orientation) or longitudinal (parallel to floodplain orientation) tilting. Both conditions are addressed in this section, with special attention to the effects of lateral tilting.

Lateral tilting

Lateral tilting is prevalent in half-graben, pull-apart, foreland, and thrust-generated sag basins (Figure 8.1). Local tilting, however, may also be caused by subsidence and/or uplift owing to fluid extraction, igneous intrusion, or evaporite intrusion and solution (Peakall, 1996). In addition, expansive "stable" areas in continental interiors may be tilted owing to transmission of intraplate stress from plate boundaries (Adams, 1980), large-scale differential compaction, or variations in isostatic recovery. In each of these examples, the site of maximum subsidence is the location to which streams will migrate (Potter, 1978; Alexander and Leeder, 1987). As an exception, some tilted basins are characterized by levels of sediment input from one margin that is sufficient to fill the site of maximum subsidence. In such situations, axial rivers shift to positions lateral to the basin tectonic axis (Figure 8.2) (Leeder and Gawthorpe, 1987; Blair and Bilodeau, 1988).

Three styles of lateral channel migration may result as a response to lateral tilting: (1) instantaneous avulsion triggered by a single tectonic event; (2) topographically triggered avulsion when a channel slowly aggrades to a point that enables a stream to steepen its profile by switching position to a new

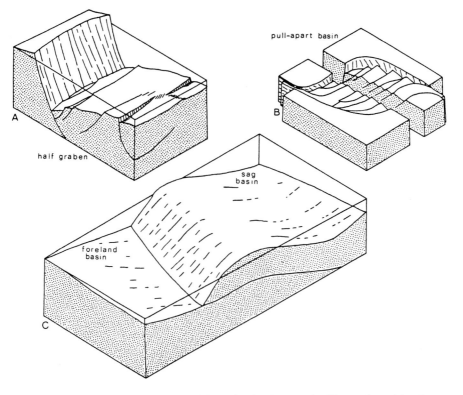

Figure 8.1 Tectonic basins which tend to be characterized by syndepositional tilting of the basin floor (from Alexander and Leeder, 1987).

steeper course, while flow continues in the channel that is being abandoned; and (3) "combing" (Todd and Went, 1991), which refers to slow migration by preferential downslope erosion and/or meander cutoff on one side of the river (Alexander and Leeder, 1987; Leeder and Gawthorpe, 1987). In general, avulsion styles 1 and 2 tend to produce isolated sand ribbons and immature channel belts, whereas combing, style 3, results in wide channel belts (Alexander and Leeder, 1987; Peakall, 1996) (Figure 8.3). If rivers comb, instead of avulse, tilting must generate gradients that are large enough to induce preferential lateral migration but small enough not to cause immediate avulsion. The limited number of quantified examples suggest that areas of lateral migration by avulsion have tilt rates of $\geq 7.5 \times 10^{-3}$ radians ka^{-1}, but examples of lateral migration by combing have tilt rates 2–3 orders of magnitude smaller (Peakall, 1996).

Although combing suggests tilting, avulsion styles 1 and 2 may occur as the result of normal river behavior or style 1 can be the result of megafloods or,

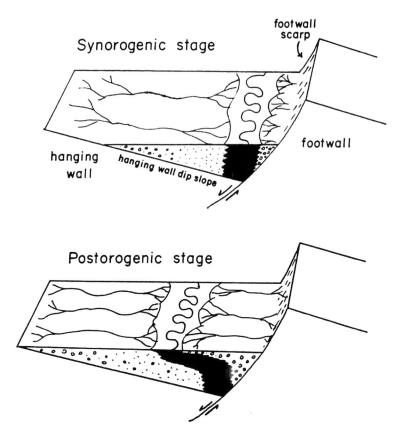

Figure 8.2 Two-stage model for fill of a tilted half-graben basin. During synorogenic time, sediment input from the footwall is projected to be low, and the axial stream is close to the basin-margin fault. During the postorogenic stage, sediment input from the footwall is projected to be high, forcing the axial stream toward the hanging wall (from Mack and Seager, 1990).

indeed, tectonics. However, it should be possible to discriminate between styles 1 and 2 in the stratigraphic record.

Fisk (1944, 1947) did some of the seminal work on filling of abandoned channels. He noted that channels that experienced neck cutoffs were mostly filled by deposition of suspended load during overbank events, and were thus composed mostly of laminated silt and clay (blue muds) (Figure 8.4), although neck cutoffs commonly had sandy plugs at the junction between the cutoff and the main stream. In contrast, he noted that chute cutoffs typically contain a larger component of coarse sediment. Active flow is maintained during early stages of cutoff filling, and the channel aggrades from the bottom and sides as flow is steadily lost to the new channel.

A key element of Fisk's discussion is that bed-load deposition in channel

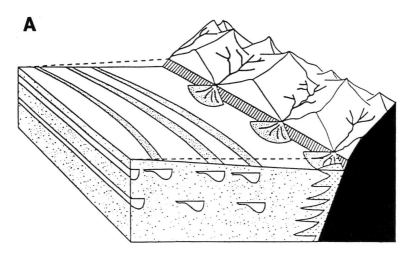

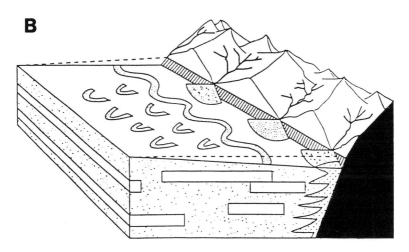

Figure 8.3 Idealized end-member models of three-dimensional fluvial architecture generated in tilted basin fills as a result of (A) down-tilt avulsion and (B) lateral combing (from Peakall, 1996).

fills is characteristic of flow maintenance during gradual abandonment, whereas deposition of finer suspended load is indicative of more abrupt and complete diversion of flow from the channel. This general principle holds whether abandonment is by cut off of a meander, or by avulsion of a longer channel reach. Special care must always be taken, however, that sandy components in channel fills represent bed-load deposition, and not sand from an overbank environment (e.g., splays, oxbow deltas, etc.), before this tenet is applied.

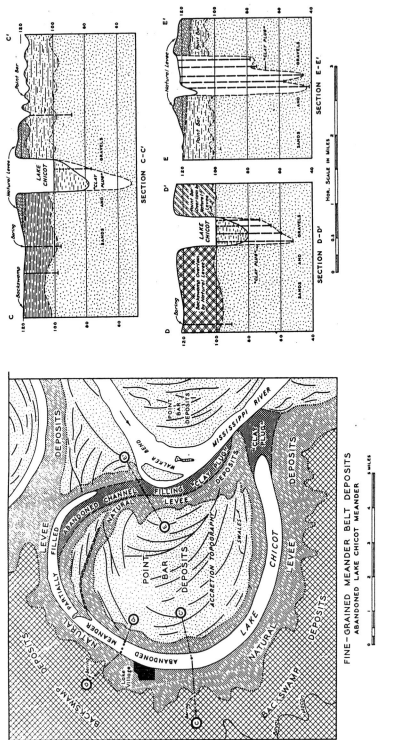

Figure 8.4 Sediments deposited in a neck-cutoff channel (from Fisk, 1944). Portions of the channel becomes filled with sands. Most of the channel, however, becomes filled with blue mud, clays, silty clays, or silty sands, and this filling is referred to locally as a "clay plug". The abandoned Lake Chicot meander shows a typical distribution of fine-grained meander-belt deposits. The natural levees are wide on the outside of the bend where they overlie backswamp accumulations and narrower and steeper on the inside of the bend where they bury the accretion topography of the point bar close to the channel. The huge point bar is made up of well defined ridges and swales that mark stages in the growth of the bend. The abandoned channel holds a large lake but is partially filled with a "clay plug". The thickness and distribution of the deposits are shown on Sections C–C', D–D', and E–E'.

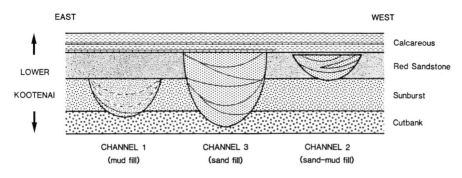

Figure 8.5 Channel fills in the Kootenai Formation along the Missouri River in Montana (from Hopkins, 1985).

Hopkins (1985) describes three channel fills of the Lower Cretaceous Kootenai Formation of Montana (Figure 8.5). These three channels are each incised into delta plain and bay-fill strata, and they are interpreted as recording avulsion of relatively straight distributary channels in association with delta lobe switching. Channel 1 (Figure 8.5) is composed of laminated mudstone and reflects filling in an overbank environment after the channel was completely and permanently separated from active flow by an abrupt avulsive event. Channel 2 was subject to intermittent occupation of the channel by active flow, probably during floods, and the channel fill alternates between beds of suspended mudstone/siltstone and bedload-originated sandstone. Channel 3 is filled with cross-bedded to parallel laminated medium-grained sandstone, and it is interpreted to represent a condition of progressive abandonment, whereby active flow and bed-load transport was maintained in the channel during the filling process. Channel 3, because of its size, may be a valley-fill deposit containing multiple sand-filled channels, although most valley-fills fine more obviously from bottom to top of the deposit (Schumm and Ethridge, 1994).

Channels undergoing gradual abandonment are subject to episodes of scouring and aggradation owing to fluctuating discharge, or "cut-and-fill", throughout the filling process. This can result in "nested" channel scours within coarse channel fills (Figure 8.6). Such "nested" scours are generally symmetrical to near-symmetrical about the channel axis, and reflect the development of smaller channels within the old avulsed channel, as discharge progressively wanes (Hopkins, 1985; Holbrook, 1996a).

Bridge and Leeder (1979), Alexander and Leeder (1987) and Gawthorpe and Colella (1990) modeled effects of tilting on alluvial architecture and noted that preferred channel avulsion toward the fault in a tilted basin will cause channel belts to be clustered on the downtilt side of the basin (Figure 8.7).

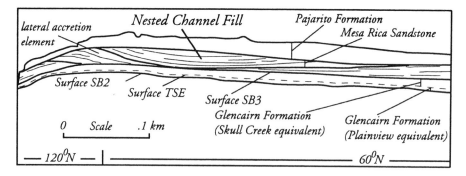

Figure 8.6 "Nested" channel scours in Lower Cretaceous fluvial sandstone of northeastern New Mexico (from Holbrook, 1996a).

Such modeling has shown that floodplain tilting does not affect cross-section averaged values of channel deposition, but it does reduce the effective flood-plain width, causing channel-belt proportion and interconnectedness to increase locally on the downtilt side (Alexander and Leeder, 1987; Leeder and Alexander, 1987; Bridge and Mackey, 1993). When the subsidence rate is greater than the migration and aggradation rate of the channel, the river will have no impetus to leave the locus of greatest subsidence, and channel belts will simply stack vertically (Todd and Went, 1991).

These models can be used to infer effects on fluvial architecture beyond channel-belt interconnectiveness. For instance, Alexander and Leeder (1987) noted that soils are exposed longer with little overbank deposition on the uptilted side of tilted floodplains, resulting in more mature paleosols. Channel-belt deposits will also tend to be wider in tilted basins if combing was a significant component of channel migration (Leeder and Alexander, 1987). In addition, avulsion-related sequences in floodplain sediments (Kraus, 1996, will tend to be thicker on the down-thrown side of the floodplain owing to proximity of the channel and greater subsidence (Bridge and Mackey, 1993).

Little confirmation of these model predictions is available. One notable example is in southern New Mexico, where Mack and James (1993) compared syntectonic Plio-Pleistocene braided fluvial deposits from asymmetrically (Palomas and Mesilla basins) and symmetrically (Hatch–Rincon and Corralitos basins) subsiding sub-basins of the Rio Grande rift system. Detailed quantification of channel elements from strata in the Palomas half-graben support the Bridge and Leeder (1979) models in that alluvial channel-belt deposits were concentrated on the down-tilt side of this tilted basin (Mack and James, 1993; Leeder *et al.*, 1996). Mack and James (1993) deter-mined that, compared to symmetrical basins, fluvial deposits in asymmetrical (tilted) basins have the following characteristics: (1) a narrower effective

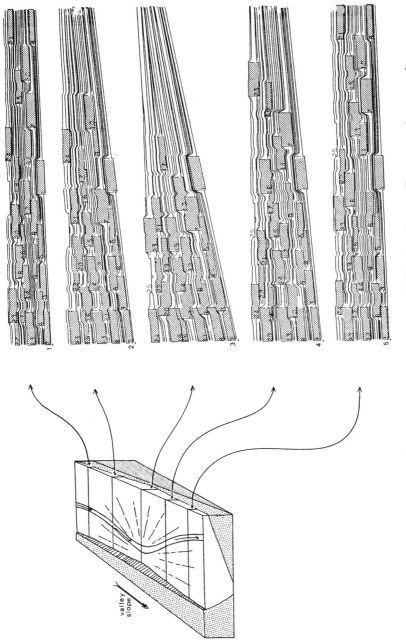

Figure 8.7 Computer model showing effect of various degrees of lateral tilting in a down-valley progression on sandstone architecture. For sections 1 to 5 the models assume relative movements for opposite sides of the floodplain to be 0.5, 2.0, 4.0, 2.0, and 0.5 meters, respectively (from Alexander and Leeder, 1987).

floodplain (where effective refers to areas of floodplain prone to channel occupation); (2) a higher percentage of multistory channel sandstone bodies; (3) a higher ratio of channel to floodplain deposits near the basin axis; (4) less paleosols and paleosols with a generally lower degree of maturity near the basin axis; and (5) a low proportion of sets of planar cross-beds. They attributed difference 1 to the tendency of the active floodplains to be restricted to a zone of maximum subsidence near the footwall block in asymmetrical basins (Mack and Seager, 1990) because this tends also to be the area of lowest topography (Bridge and Leeder, 1979; Alexander and Leeder, 1987; Leeder and Alexander, 1987; Leeder and Gawthorpe, 1987). They postulated that, because the effective floodplain is narrow in an asymmetric basin, laterally migrating braided channels return to old channel sites, before a thickness of floodplain sediment can accumulate over the channel deposit. This tends to promote channel-belt amalgamation and development of multistory channel sandstone (difference 2), while limiting the likelihood of floodplain mud preservation (difference 3) near the basin axis. The short duration between channel occupation for any part of the effective floodplain also tends to decrease the likelihood for mature soil development (difference 4) in such areas. They suggested that the paucity of planar cross-sets in asymmetric basins might reflect the low survivability potential of these structures because of the greater degree of channel reworking in these basins (differences). This idea, however, remains untested.

For meandering rivers, down-tilt combing tends to not only affect the geometry and spatial distribution of channel belts (i.e., wider channel belts and channel belts concentrated on the down-tilt side of the basin) but also the internal architecture of channel-belt deposits. In studies of the tilted floodplain of the Madison River, Montana, Leeder and Alexander (1987) noted that preferential meander-belt shift in the down-tilt direction tends to destroy all cutoff loops generated on the downslope side of the meander belt (Figure 8.8). There are three important implications from this observation that apply to channel-belt architecture in fluvial sediments of syndepositionally tilted basins: (1) lateral accretion surfaces will tend to dip preferentially in the up-tilt direction; (2) meander-belt sand bodies without preferential dip or lateral accretion surfaces can be assumed to have evolved randomly without the influence of imposed tectonic slope; and (3) under aggrading conditions, the base of the sand body produced by combing will climb up-section in the down-tilt direction (Leeder and Alexander, 1987).

Further examination of the Madison River by Alexander *et al.* (1994) revealed additional details of the combing process and resultant sediments. They confirmed Leeder and Alexander's (1987) assertion that the Madison

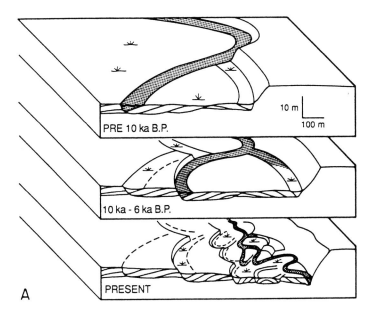

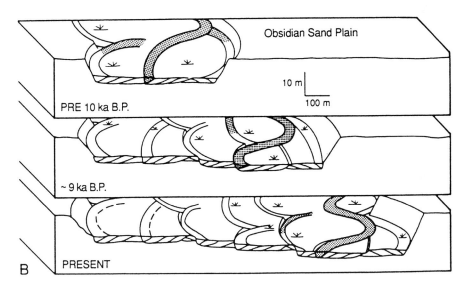

Figure 8.8 Examples of lateral combing by the South Fork (A) and the Madison River, Montana (B), producing localized channel-belt incision and lateral accretion surfaces that dip asympathetically with tectonic tilt direction (modified after Alexander *et al.*, 1994).

River shifted laterally toward the footwall of the fault-tilted graben over time, producing asymmetric meander belts with up-dip slanted lateral-accretion bends and abnormally wide coarse sediment bodies. Superimposed on this combing, however, are several episodes of incision and aggradation of individual channel belts; perhaps the result of climatic, tectonic, or autocyclic controls. Some younger belts were incised below older belts, complicating the picture of a simple progressive lateral shifting of belts over time (Figure 8.8). They noted, however, that, despite climate effects, there was still an apparent general trend for preferential migration of the river in the down-tilt (eastward) direction, as based on radiometric age of meander loops. However, they warn that "we urge caution and the use of interdisciplinary studies before geologically ancient asymmetric meander-belt deposition is interpreted as purely of tectonic origin."

Nanson (1980a) observed an exception to the Madison River example from the tilted floodplain of the Beatton River, Canada. Here, meander loops were cut off preferentially on the down-tilt side of the river channel because the amplitude of these loops was greater than the amplitude of the up-tilt bends. This appears to be a special case, however, where the Beatton River was confined to a narrow valley and unable to comb freely (Peakall, 1996).

Few studies thus far have attempted to verify the Leeder and Alexander's (1987) and Alexander et al. (1994) observations on tilting and preferential lateral-accretion surface dip in more ancient meandering river deposits. One example is provided by Hazel (1994) who attributed architectural facies variability in the Upper Triassic Chinle Formation to localized tilting of the Moab salt anticline, within the Paradox Basin, southeastern Utah. Meandering channel deposits of the Kane Springs member of the Chinle Formation (sand sheet #2) have a predominance of lateral accretion surfaces pointed northeast toward the Moab anticline axis and away from the southwestward flanking King's Bottom syncline axis. Hazel (1994) attributed this to preferential preservation of up-tilt meander belts due to the down-dip migration of meandering streams on a tilting Moab anticline fold limb. Angular intraformational unconformities in Chinle strata independently attest to salt uplift and fold-limb tilting on the Moab anticline during Kane Springs deposition (Hazel, 1994). Woolfe (1992) also noted a similar preservational bias in lateral-accretion surface orientation in the Weller coal measures of Allan Hills, Antarctica, that he attributed to the tilt-generated combing predicted by Leeder and Alexander (1987).

Signs of lateral combing may also be preserved in the internal architecture of braided channel deposits. Todd and Went (1991) studied the Devonian Glashabeg Formation and the Cambrian Alderney Sandstone Formation and identified two architectural features, which may have been generated as low-sinuosity braided-river channels which migrated preferentially in one direction. The first feature is asympathically dipping (dipping in direction opposite of channel combing) planar cross-bed sets, which are characteristic of the Glashabeg Formation. In this case, lateral bars were detached from banks during floods by a chute channel. During times of flooding, waters rushed across the bars, causing sand avalanches and planar cross-bed accretion on the inner bank of the chute channel. These deposits appear as trough-shaped channel scours filled by planar cross-beds that are dipping 90° to scour orientation. Combing of channels results in preferential preservation of lateral bars and chute channel fills on the side of the channel opposite the direction of combing, thus the asympathic cross-bed sets are preferentially preserved.

The second architectural feature is sympathic dipping planar cross-bed sets. These characterize the Alderney Sandstone. Lateral bars in Alderney channels grew as planar cross-sets that accreted toward the channel center and covered the thinly interbedded mudstone and sandstone lamina, that tended to drape the channel floors between floods. These cross-bed sets were capped by parallel laminated sandstone beds that were deposited during shallow flow over bar tops. Preferential preservation of lateral bars on the side of the channel opposite the combing direction means that planar cross-bed sets in this association will preferentially dip sympathically with the direction of channel migration. Todd and Went (1991) stressed that a preponderance of sympathic or asympathic planar cross-beds in braided stream deposits only reflects lateral combing. They felt, however, that lateral migration of Glashabeg and Alderney channels was driven by tectonic tilting. They further stressed that planar cross-bed sets in these low–medium sinuosity sandy to gravelly streams tended to represent bar accretion, which could be up to 90° from the true paleoflow direction. These vectors thus tended to reveal the tilting bias, whereas the trough cross-bed sets represented more in-channel flow, and was more reflective of true paleocurrent direction.

Paleocurrent orientations have also been used to argue for tilting and lateral migration of fluvial channels. Martins-Neto (1994) observed that Proterozoic fluvial strata from axial rift-fill deposits of the São João de Chapada Formation, Brazil, shift from east to southeast upward in the succession. He used this to argue for progressive tilting and lateral migration of channels southward (perpendicular to the basin axis) over the course of fluvial deposition. In another study, Nakayama (1994) developed tilting histories for three sedimentary intra-arc basins in Japan (Tokai, Kobiwako, and Osaka groups), as based on geologic structures and thickness of strata between widespread volcanic beds. Paleocurrents from axial streams (flowing parallel to fault strike) show a strong paleocurrent bias in the direction of tilting (normal to channel orientation) for intervals independently verified as reflecting deposition in a tilted basin. In addition, Nakayama (1996) observed that deposits of axial middle- to low-sinuosity braided streams corresponded to times of unidirectional tilting, but mud-rich high-sinuosity streams corresponded to times of changing tilt direction in these Japanese intra-arc basins. No explanation was given for this phenomenon.

Longitudinal tilting

Discharge, sediment load, and slope are the primary controls on channel pattern (Lane, 1957; Leopold and Wolman, 1957; Schumm, 1968). Sufficient changes in slope generated by down-dip tilt will thus result in changes in channel pattern (see Chapters 2, 3, and 4), which may be recorded in sedimentary deposits. For instance, increased downstream tilting can increase stream sinuosity and, if large enough, induce braiding (Schumm, 1977, 1986; Adams, 1980). If other factors (e.g., discharge and sediment load) can be eliminated as a cause, down-slope shifts in fluvial architecture that are reflective of a shift in stream pattern may be attributed to longitudinal floodplain tilting.

Mackey and Bridge (1995) modeled migration of avulsion nodes, and noted that they have a tendency to migrate up valley over time if no longitudinal tilting occurs. In fact, in back-tilted basins, almost all the channel belts are produced by upstream avulsion. This is caused by a natural increase in aggradation at the avulsion node and slope decrease in the down-dip direction as sedimentation continues. By inference, if the rate of tectonic subsidence increases downslope, up-dip node migration becomes less likely.

Sufficient longitudinal tilting may even result in paleoslope reversal. Fluvial braided and high sinuosity deposits of the Lower Devonian Battery Point Formation record such a shift. Paleocurrent data from this unit indicate that westerly directed axial transport in the subsiding Gaspé trough (Canadian Maritime provinces) shifted to east and southeast axial flow as impingement of the Avalon terrain against the New York Promontory forced uplift of the western end of the floodplain during the Devonian Acadian Orogeny (Lawrence and Williams, 1987).

Effects of subsidence

Probably the most important effect of subsidence on fluvial sedimentation is the generation of space for fluvial deposition through lowering of the basin floor. Allen (1978) and Bridge and Leeder (1979) are often credited with first quantifying the interrelationship between subsidence and large-scale fluvial architecture (the original LAB models). They stressed that interconnectivity of channel belts (*sensu* Allen, 1965) in an alluvial suite was highly dependent on subsidence rate. In their models, they assumed that a channel will remain in a set part of the floodplain, where it produces a channel belt and alluvial ridge. At some time later the stream avulses to a random (Allen, 1978) or topographically lower (Bridge and Leeder, 1979) position on the floodplain. Based on modern analogy, the avulsions occured at 10^2–10^3 yr. Model runs held the avulsion rate constant, but allowed the subsidence rate to vary through successive iterations. The models also assumed that channels maintained grade by filling accommodation space generated by subsidence, and that areas not filled by sand-rich channel belts were filled by clay-rich overbank deposits.

The fundamental result of these models was that avulsion under conditions of slow subsidence promoted prolific incisement and reworking of overbank fines by channels. This caused high interconnectiveness of channel belts and deposition of multi-story and multi-lateral sandstone sheets. In contrast, rapid subsidence permitted significant deposition of overbank fines before reinitiation of channel belt deposition at a new location, thus promoting low channel-belt connectivity (Figure 8.9). Several studies have since utilized these concepts, citing high interconnectivity of channel belts, and resultant high sandstone/mudstone ratio, as evidence of slow subsidence rates, and vice versa (Bridge and Diemer, 1983; Blakey and Gubatosa, 1984; Kraus and Middleton, 1987; Nichols, 1987; Shuster and Steidtman, 1987; Clemente and Pérez-Arlucea, 1993).

Sandstone/mudstone ratios have also been used as an indicator of channel pattern, with high sandstone/mudstone ratios commonly associated with braided channels and low sandstone/mudstone ratios related to meandering channels (Bridge, 1985). Pattern is related to slope, sediment supply and discharge, and it is not necessarily related directly to subsidence rate. This must be considered before inferences of subsidence rates are imposed on fluvial deposits based on sandstone/mudstone ratios. Because an upward shift in sandstone/mudstone ratio could represent either a change in subsidence rate or a change in stream pattern, direct application of the Allen (1978) and Bridge and Leeder (1979) models to explain such shifts as tectonic in origin requires that depositional features likely to be altered by changes in fluvial pattern be compared for similarity vertically throughout the section (e.g., dominant bar forms, proportion of channel and lateral-accretion deposits, lithology and structure of channel fills, paleocurrent variance, channel and bar direction/geometry, etc.; Bridge, 1985). Architectural element analysis (Miall, 1985, 1991) is an invaluable tool for such comparisons.

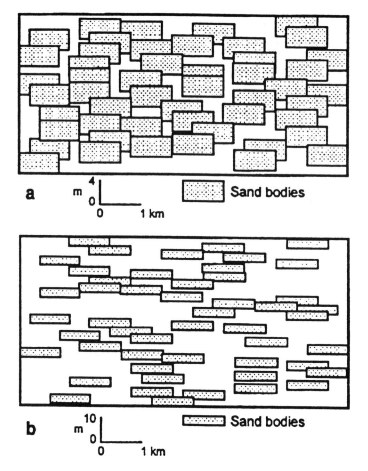

Figure 8.9 Modeled interconnectiveness of channel belts deposited under conditions of (a) low subsidence rate and (b) rapid subsidence rate (from Allen, 1978).

The Allen (1978) and Bridge and Leeder (1979) models have recently experienced refinements and criticisms. However, some support exists for acceleration of avulsion rates with increased subsidence rate. Bryant *et al.* (1995) conducted flume experiments to test controls on avulsion rate, and found that avulsion rate accelerated as the sediment feed rate was increased to experimental streams. Törnquist (1994) also showed that avulsion rates increased in channels of the Rhine–Meuse delta synchronously with, and probably related to, generation of increased accommodation space by currently rising sea level. Additionally, Read and Dean (1982) found relationships between sandstone/mudstone ratio and aggradation rates that appeared opposite of those predicted by Allen (1978) and Bridge and Leeder (1979) in Carboniferous fluvial deposits of Scotland.

Bryant *et al.* (1995) and Heller and Paola (1996) note that there is serious lack of understanding of the relationship of avulsion rate to aggradation rate, and that flume and quantitative models can not be directly extrapolated to sedimentary deposits. Likewise, models and experiments that require equating sedimentation rates to subsidence rates, may not be valid. Their point is well taken, however, that caution should be used when applying the Allen (1978) and Bridge and Leeder (1979) models until some resolution is reached on the issue of subsidence and avulsion rates.

It is worth reiterating in this section that sea level rise or stillstand progradation can generate accommodation space similarly to subsidence. Slowly rising or stable sea level can result in slow generation of accommodation space and intensive channel belt amalgamation, in fluvial systems proximal to the strandline (Shanley and McCabe, 1991; Holbrook, 1996a). In such cases, distinction of tectonic or eustatic control on channel-belt interconnectivity may prove difficult to impossible. In such coastal areas, "relative sea level" behavior may be all that can be assessed from channel belt interconnectivity.

Effects of uplift

Syndepositional uplift produces various effects on alluvial channels that can be preserved in the stratigraphic record. Uplifts within a depositional basin may control lithofacies distribution as determined by stream course or pattern. Uplifts outside a depositional basin (extrabasinal uplifts) are subject to erosion, and often serve as the primary source of sediments for basin fill. Numerous provenance studies have used mineralogy of basin sediments to derive locus, climate, and bedrock geology of such uplifted sediment source areas (Folk, 1974; Basu *et al.*, 1975; Dickinson and Suczek, 1979; Lloyd, 1994; Espejo and López-Gamudí, 1994). The following section, however, concentrates on the propensity of extrabasinal uplifts to generate coarse sediments within fluvial depositional systems.

Extrabasinal uplift

Conceptual ties between coarse-grained sedimentation and uplift derive from the relationship between uplift and gravel generation determined by Playfair (1802) and Barrell (1917). In summary, increased slope results in increased energy in uplifted rivers. This increases the competency of rivers, and permits transport and basin delivery of larger grains. Source-area uplift thus tends to result in gravel progradation in the basin. It is important to note, however, that basin subsidence may produce relative relief between source area and basin as well, having the same effect of increased slope and energy without an actual increase in source-area elevation.

Progradation of coarse clastic sediments in a basin is traditionally argued to be coincident with uplift of the source area, inferring a direct linkage between age of conglomeratic strata and timing of source-area uplift (Steel *et al.*, 1977; Rust and Koster, 1984). Although not disputing the importance of uplift in gravel generation, several authors have forwarded the idea that source-area uplift and basinal gravel progradation are not always synchronous, owing to the effects of basin subsidence and erosional lag time on gravel generation and progradation. An important factor in common to each of these studies is that gravel progradation will fundamentally be controlled by the ability of the basin to trap sediments near the basin margin through rapid generation of accommodation space by subsidence (Blair and Bilodeau, 1988; Flemings and Jordan, 1989; Heller *et al.*, 1988; Paola, 1988; Gordon and Heller, 1993).

Beck (1985), Beck *et al.* (1988), and Beck and Vondra (1985) were some of the first workers to notice this discrepancy between timing of uplift and gravel progradation. They observed that conglomerate in Rocky Mountain foreland basins that was deposited during episodes of thrusting is close to the sourcing thrust, whereas the conglomerate sheets that prograded across the basins are primarily postorogenic.

Blair (1987) studied alternating gravel (basin margin facies) and fine (basin axial facies) strata in the non-marine Upper Jurassic/Lower Cretaceous Todos Santos Formation of Chiapas Mexico. He concluded that times when axial facies were near the basin footwall margin, and thus times of minimal gravel progradation, represent the times of rapid subsidence, generation of higher relative relief between source area and basin, and trapping of gravels near the basin margin. In contrast, times of gravel progradation and displacement of axial environments reflect a postorogenic phase (Figure 8.2).

Walker River and Walker Lake in the Walker Lake half graben in Nevada serves as an example of rapid response of basin axial environments to tectonics. Human river diversion, beginning in 1882, has resulted in lowering of Walker Lake, and exposure of the northern part of Walker Lake to river occupation. Historical records indicate that Walker river began lateral migration as a response to tectonic tilting of the half graben in a matter of a few years, to a few decades, after occupying this newly exposed area (Blair and McPherson, 1994).

Blair and Bilodeau (1988) and Blair and McPherson (1994) assert that, from the perspective of geologic time, rivers and lakes can respond almost instantaneously to tectonic stimuli. Likewise, they assert that, because rates of tectonic uplift on average are eight times higher than average denudation rates (Schumm, 1963), a lag time will exist between uplift and gravel deposition in a subsiding basin. Accordingly, they argue that gravel progradation in a subsiding basin will tend in general to occur well after major episodes of subsidence and relative source-area uplift (Postorogenic stage; Figure 8.2). Times when greater relief is generated between the source area and a subsiding basin will be marked by deposition of finer basin-axial facies near the source area (Synorogenic stage; Figure 8.2).

Other studies have produced similar results. Gordon and Heller (1993), for instance, examined times of major basin subsidence and gravel progradation from late Miocene to Recent in the Pine Valley half graben of Nevada, and found that times of major progradation correspond to episodes of low subsidence. Mack and Seager (1990) found similar results in the Plio-Pleistocene Camp Rice and Palomas Formations of the southern Rio Grande Rift. Cant and Stockmal (1993) found that these relationships also held for the Mesozoic fill of the Alberta foreland basin. They further noted from their stratigraphic and structural data that the lag time between initial thrusting and gravel progradation is longer than that of subsequent thrusting, possibly owing to the relatively continuous background supply of orogen-derived sediment once the coupled basin and thrust system is established.

Heller *et al.* (1988) echoed the findings of Blair (1987) in his studies of the U.S. Cretaceous Western Interior Basin by inferring that gravels which prograded well distal of the thrust front post-dated timing of major thrust advance. They added a twist, however, by speculating that flexural rebound of the foreland basin deposits ensued as flanking thrust sheets lost mass through erosion. This could have caused uplift and erosion of recently deposited gravelly sediments near the thrust front prompting gravel progradation.

Burbank *et al.* (1988) dispute usage of the postorogenic model for gravel progradation in the Himalayan foreland basin. Here, high-resolution magnetostratigraphy indicates that gravel propagation in the Himalayan front is virtually synchronous with thrusting. Gravels had extended more than 110 km into the basin by 3 Ma after the start of thrusting. They attribute the difference to the abnormally high rates of erosion and transport from the source area, which fills available space in the subsiding basin, forcing overspilling of the basin and progradation of conglomerate during thrusting. They note that where such factors are at play, syntectonic gravel progradation may occur.

Intrabasinal uplift

Intrabasinal uplift here refers to relative highs generated within a depositional basin. These uplifts may represent areas of active tectonic uplift, or simply areas of reduced subsidence, that produce topographic highs relative to surrounding depositional systems. Such uplifts may have sufficient elevation to undergo erosion, although the designation "intrabasinal" infers that these uplifts are not likely to have sufficient elevation to serve as a volumetrically significant sediment source area. In most cases discussed here, uplifts lack sufficient elevation to undergo major erosion, but they still have sufficient elevation to influence depositional patterns.

Though intrabasinal uplift may result from igneous or salt intrusion, deformation owing to intrabasinal tectonic stresses is probably the more common causative mechanism. Such intrabasinal stresses are weak, and most typically reactivate pre-existing structures rather than break new crust (Sykes, 1978). Several examples of uplift through reactivation of pre-existing structures are noted in the literature (e.g., Holbrook and Wright Dunbar, 1992; Meyers, et al., 1992, Lorenz, 1982). Marshak and Paulson (1996), in fact, conclude that almost all of the recently active mid-continent fault and fold zones in the U.S. result from reactivation of Precambrian faults. Holbrook (1996b) notes that intraplate stresses will generally be sufficient to prompt generation of topographically low uplifts ($\leq 10\,m$) by structural reactivation throughout continental interiors, irrespective of the coincidence of a major orogeny. These low uplifts can thus be considered a form of "structural noise" that may appear within any basin or plate interior at any time. For example, see discussion of the Mississippi River Valley in Chapter 9.

Differential compaction may produce relative intrabasinal uplifts of non-tectonic origin (Gay, 1994). For instance, le Roux (1994) cites three examples (the Nours Formation, Australia; the Ramon and Arad Groups, Israel; and the Beaufort Group, South Africa) where marine and terrestrial depositional trends are consistent with depositional trends from previously deposited strata, yet no tectonic structures are apparent which could have influenced these trends over long periods. These trends, however, appear too similar between units to be coincidental. The common denominator in each of these three cases seems to be the presence of thick argillites subjacent to denser sandstone or evaporites. Water expulsion and differential compaction under the weight of these denser deposits is blamed for creating sinks that served as receptacles for younger sediments, and promoted propagation of depositional trends to overlying units.

Fluvial stratigraphy

Intrabasinal uplifts may be manifest in the stratigraphic record in several ways. Erosional intraformational angular unconformities and localized thinning of fluvial strata above uplifts are such indicators of syndepositional uplift (Miall, 1978). Anadón *et al.* (1986) and Riba (1976) in particular point out the importance of recognizing intraformational unconformities for spotting syndepositional uplifts. Several cases for intrabasinal uplift have been made based on development of intraformational angular unconformities and localized thinning trends in fluvial strata. Schwartz (1982), for example, argued that Early Cretaceous reactivation of the Boulder Batholith high within the Western Interior foreland basin of southwestern Montana was apparently sufficient to generate radial paleocurrent patterns in fluvial sandstone flanking the uplift, thinning of fluvial strata above the uplift (DeCelles, 1986), and increase in the lithic component of flanking sandstone bodies, presumably due to intraformational erosion over the uplift as an intraformational angular unconformity was generated (Schwartz, 1982).

Similar conclusions were reached by Kirk and Condon (1986), that syndepositional deformation during the late Jurassic controlled the distribution of sedimentary facies in the Westwater Canyon and Brushy Basin members of the Morrison Formation, which controlled the localization of uranium ore deposits. Rising structures during deposition resulted in (1) low total thickness of the formation, (2) low sandstone/mudstone ratios, (3) low sandstone percentages, (4) greater number of mudstone beds over the uplift. The reverse situation existed in depocenter axes. In addition, Guiseppe and Heller (1998) determined that the alluvial architecture of the Price River formation changed from mostly sheetlike sandstones to isolated sandstone lenses as the area affected by the active swell was approached by the ancient rivers.

Meyers *et al.* (1992) noted that conglomerate in the Lower Cretaceous Cloverly Formation of Wyoming existed everywhere but above the currently up-thrown fault blocks of the Laramide Wind River Range. They argued that Early Cretaceous uplift on these blocks caused conglomeratic sandstone to be reworked and stripped from uplifted segments of Wind River blocks. Similar effects can be observed in deposits elsewhere in the world (Krzyszkowski and Stachura, 1997). For instance, syndepositional tectonics resulted in faulting and folding of sediments in the Pliocene–Pleistocene upper Siwalik Group of the Himalayan foreland basin in Pakistan, resulting in intraformational angular unconformities and faults over basement block uplifts that partitioned the foreland during filling. Uplift of these blocks eventually produced conglomerates flanking the partitioning uplifts (Pivnik and Johnson, 1995). Mass flow deposits adjacent to structures within the Pine Creek Sandstone, thinning of the Corbin Sandstone over the Rockcastle uplift, and deflection of paleocurrents coincident with the Irvine–Paint Creek fault can also be tied to local tectonic influence on Lower Pennsylvanian braided streams in the Appalachian Basin (Greb and Chesnut, 1996).

River diversions

Intrabasinal uplifts quite typically are very subtle, and do not have sufficient relief to yield sediments or produce obvious intraformational angular unconformities. Such uplifts may still, however, possess sufficient elevation to affect stream pattern or position through gradient change as discussed earlier in this chapter and in Chapters 3 and 4. Streams encountering such intrabasinal uplifts on their down-valley journey are faced with two choices, go around these uplifts, or go over them. Either choice will probably alter the stream in ways that are preserved in and detectable from the stratigraphic record.

Numerous cases of stream diversion by subtle intrabasinal uplift are recognized in the stratigraphic record. Many of these cases are recognized by deflection of paleocurrent trends away from sites of repeated structural uplift, or variations in occurrence of channel-belt lithofacies coincident with structures. For example, Pliocene–Recent fluvial deposits of the Sorbas Basin of southeast Spain have paleocurrent trends that show deflection around sites of modern uplift and into flanking structural lows (Mather, 1993). Syndepositional elevation of these uplifts is further evidenced by the presence of intraformational angular unconformities above these structural highs (Mather, 1993). Sand isoliths in the Lower Cretaceous Antlers Formation of central Texas reveal a preferential increase in sand thickness (as much as 2× as much sand) in pre-Cretaceous structural lows (i.e., the Kingston and Sherman synclines), suggesting Cretaceous reactivation and sagging of these features and preferential diversion of sand-depositing river channels into these coevally produced tectonic lows.

Triassic salt tectonics in southeastern Utah caused diversion of contemporary streams around salt anticlines in a manner comparable to structural diversions discussed above. This is evidenced by paleocurrents from fluvial sandstone in the Triassic Chinle formation that parallel the Moab and Cane Creek salt anticline axis (Hazel, 1994). Channel belts also tend to be concentrated in the King's Bottom syncline, between the Cane Creek and Moab anticlines, and they show decreased interconnectiveness and complex interfingering with mud over the Cane Creek crest (Hazel, 1994). This suggests that channel positioning, and thus channel-belt deposition, rarely occurred over anticlines. Chinle stream patterns also apparently changed as fluvial aggradation buried salt anticlines. In the Kane Springs member (basal sheet sandstone), coarse-grained meander belts with lateral-accretion surfaces and point-bar sequences characterize the bottom of the unit in locations of structural lows (Hazel, 1991). Such point-bar features are absent throughout the upper part of the member, and these strata are interpreted to reflect deposition in unconfined braided channels that completely buried positive areas (Blakey and Gubitosa, 1984; Hazel, 1991). Hazel (1994) interpreted this to reflect confinement and meandering of channels deflected to areas between uplifts, followed by deposition of more broadly mobile braided streams, as aggradation buried uplifts and streams became unconfined. The Kane Springs member thins over anticlines (Hazel, 1991), as would be expected in such a model.

Where rivers are degrading, diversion of rivers around structural highs may result in preferential incision of paleovalleys coincident with structurally low areas. Paleovalley trends are recognizable in the stratigraphic record, and their reconstruction can prove invaluable for identifying episodes of structural uplift on structures. Such coincidence of paleovalleys with traditionally low basement structures has been recorded by several authors (Weimer *et al.*, 1982; Weimer, 1984; Dolson *et al.*, 1991; Holbrook, 1992; Cant and Abrahamson, 1996). Weimer (1984) presents a particularly well-cited example of structurally controlled paleovalleys. In this example, close association of paleovalleys in the Lower Cretaceous Muddy Sandstone with areas between well-established historical basement highs led Weimer (1984) to assert that Early Cretaceous uplift of these structural elements resulted in deflection of rivers and preferential incision of valleys. Paleovalley trends were mapped using isopach maps of valley-fill strata that were assessed as that part of the Muddy Sandstone section between an overlying transgressive surface of erosion and a basal sequence-bounding unconformity, and by identifying locations of maximum incision, as marked by maximum truncation of marker units in marine Skull Creek shales underlying Muddy valley-fill strata (Figure 8.10).

Kvale and Vondra (1993) similarly noted control of Cretaceous valley position and shape coincident with modern structural lineaments in the Bighorn Mountains of Wyoming. Movement along lineaments of the Bighorn Mountains forced stream deposits of the Lower Cretaceous Himes member of the Cloverly Formation to narrow and become constrained to two locations parallel to Tongue River and Bighorn/Prior boundary lineaments over the region of the current Bighorn Mountains. Away from the mountains and into the basin, correlative river deposits spread out and were not constrained. Kvale and Vondra (1993) interpreted this to represent confinement of streams into narrow valleys along these two active lineaments where streams passed over active elements of proto-Bighorn Mountains, and free dispersal of channel deposits in less constrained valley systems where streams exited the higher topography and became unconfined in the Bighorn Basin area. This interpretation is supported by fanning paleocurrents patterns in fluvial Cloverly sandstone of the Bighorn Basin.

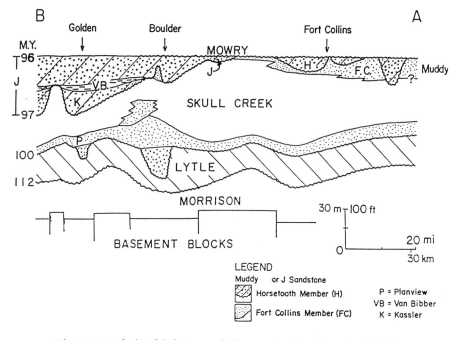

Figure 8.10 Relationship between thickness of and location of valley fills (Horsetooth Member) and uplifted basement blocks. Note incision and local removal of Fort Collins Member (modified after Weimer, 1984).

Soil development has potential for aiding in location of sites where streams were diverted around paleostructural highs as well. Srivastava *et al.* (1994) studied neotectonics of the Gangetic plains of northeastern India, and noted a trend whereby soils on slightly uplifted basement blocks were developed to a greater degree than soils on the down-dropped blocks. They attributed this to greater rates of sedimentation in the down-dropped blocks where the Ramaganga and Rapti rivers and tributaries were deflected and rates of deposition were accordingly higher. Higher areas are sheltered from deposition, and soils are allowed longer periods of stability, which enhances soil development (Srivastava *et al.*, 1994). However, where uplift is more significant, erosion typically strips soils from the uplift crests (e.g., the Sarda-Ghaghra block).

In some cases, streams may pass over uplifts rather than experience deflection. Several examples where modern stream characteristics have been affected as streams pass over the top of uplifts are cited in Chapter 9. Few examples of such effects, however, are cited from the stratigraphic record (Peterson, 1984; DeCelles, 1986; Srivastava *et al.*, 1994; Holbrook and White, 1999). Peterson (1984) noticed north to north-northwest thinning trends in the Upper Jurassic Morrison Formation of the western Colorado Plateau that coincided closely with basement and later Tertiary Laramide structures.

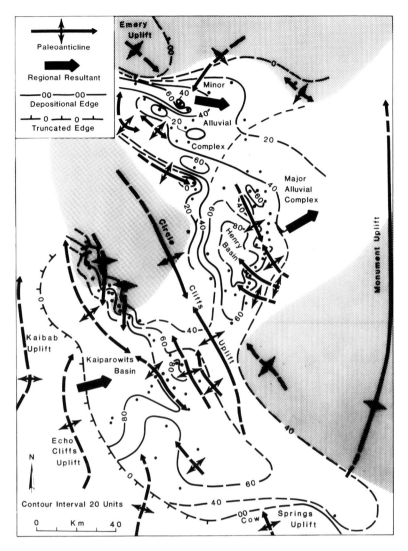

Figure 8.11 Percent sandstone in beds 0.91 m or more thick in lower Morrison Formation on western Colorado Plateau illustrating relationship between sandstone thickness, basement-cored uplifts, and regional paleocurrent trend (from Peterson, 1984).

Paleocurrents in these largely braided rivers passed over these structures perpendicular to the structural axis. Peterson (1984) noticed that relative flow regime, as suggested by stratification ratios, increased in downwarped areas and decreased over positive areas. He attributed this to shallowing of the streams in downwarped areas by lowering of gradients. He also noticed that greater quantities of sand were deposited in downwarped areas (Figure 8.11). He asserted that this was largely because lowered slopes over positive areas minimized the ability of streams to rework sediments and enhanced mud preservation in these areas.

DeCelles (1986) contended that Early Cretaceous activation of the Snowcrest Uplift in southwestern Montana apparently dammed traversing rivers of the "Second Sandstone" of the Lower Cretaceous Kootenai Formation, causing upstream reaches of rivers to become relatively straighter and possibly to take on anastomosing patterns. DeCelles' (1986) argument is based on the apparently lower sinuosity of streams above the Snowcrest axis. Low sinuosity estimates are here based on low width/thickness ratios (typical of anastomosing systems; Smith, 1983 and Rust and Legun, 1983); low variability of paleocurrent directions (suggestive of minimal meandering e.g., Miall, 1976; Schumm, 1972); and estimates of low channel sinuosity from epsilon cross-sets.

Lower Cretaceous rocks of the Mesa Rica Sandstone in northeastern New Mexico appear to have been deposited by very low sinuosity single-channel streams on the up-stream flank of the Sierra Grande basement uplift and also in areas well down paleodip of the uplift. Evidence for this is provided here by dominance of sand-dominated/active channel-fill elements over all other elements, low width/depth ratio (8–12) of channel-fill elements, an almost complete lack of lateral accretion elements, and lack of multi-lateral channel scours and multiple bar forms within major channel fills. In contrast, Mesa Rica deposits on the downstream flank of the Sierra Grande structure have approximately equal proportions of channel-fill and lateral-accretion elements and a predominance of muddy/abandoned channel fills over sandy active channel fills, which is evidence for higher sinuosity for streams in this reach. This is closely analogous to the situation described in Chapter 9 (Figure 9.10) whereby the Mississippi River is forced to increase sinuosity, as a response to steepened slopes on the downstream flank of the rising Lake County uplift (Russ, 1982; Schumm *et al.*, 1994).

Discussion

Tectonic perturbations are preserved in stream patterns and fluvial deposits. Subsidence will typically be manifest as variation in channel-belt stacking patterns, with less channel-belt interconnectivity being manifest where subsidence rates were relatively high, and vice versa. Though still utilized, this relationship has been questioned as of late, and should be used with skepticism. Likewise, relative sea level rise may result in similar associations, making linkage of channel-belt stacking patterns solely to subsidence suspect in deposits of nearshore areas.

Lithofacies and architectural patterns characteristic of preferential shifting of a channel toward one side of the flood basin, may be attributed to lateral tilting. Such patterns might include: preferential orientation of lateral accretion surfaces, preferential stacking of channel belts on one side of the flood basin, down-valley paleocurrent orientations skewed toward one side of the flood basin, and more mature paleosols on one side of a paleovalley. Longitudinal valley tilting may change valley slope sufficiently to alter stream patterns, which may be preserved as vertical or lateral shifts in architectural element association. Also, longitudinal tilting may affect avulsion node migration, which can affect channel-belt stacking patterns.

The primary effect of extrabasinal uplift on fluvial systems is in the generation of coarse clastic sediments. In general, a lag time will exist between the episode of uplift, and propagation of coarse clastics into the basin. In extreme cases, where the potential of the source area to deliver sediment is great compared to the ability of the basin to accommodate these arriving sediments (e.g., the Himalayan Front), this lag time may be negligible. The main fluvial effects preserved because of intrabasinal uplift are preferential channel-belt and valley preservation adjacent to uplifts, owing to diversion of rivers around tectonic highs, and the variation of fluvial architecture above uplifts, owing to shifts in stream pattern.

Identification of active and neotectonic structures

Previous chapters demonstrate how deformation affects alluvial rivers and their sedimentary deposits. In this chapter, what is known about alluvial river response is applied in an attempt to identify active structural or neotectonic features affecting the Mississippi River and its alluvial valley, the Atlantic Coastal Plain, and the Sacramento and Nile Rivers. The use of a hydraulic model to locate faulting in the Alabama River valley concludes the chapter. This is the inverse approach of Willemin and Knuepfer (1994).

Mississippi River valley

The Mississippi River and other rivers located in the Mississippi River alluvial valley were studied between Hickman, Kentucky and Osceola, Arkansas (Figure 9.1) in order to determine if there is evidence of active deformation in this area (Boyd and Schumm, 1995; Schumm and Spitz, 1997; Spitz and Schumm, 1997) and if there is geomorphic evidence for the existence of some postulated faults and folds. These studies were part of a major effort to develop an understanding of the New Madrid seismic zone (Figure 9.2), which is a region of great national concern in the United States due to the historic occurrence of severe earthquakes in the region (see Chapter 5). Other geomorphic studies related to the New Madrid seismic zone include identification of a deformed river profile near New Madrid (Russ, 1982), deformed terraces of streams in western Tennessee (Saucier, 1987), the formation of sunklands due to deformation in northeastern Arkansas (Guccione *et al.*, 1993, 1994), deformed stream networks (Merritts and Hesterberg, 1994) and stream profiles (McKeown *et al.*, 1988).

The New Madrid seismic zone has been studied by numerous workers since the early 1970s (Johnston and Shedlock, 1992), and much of the work that has

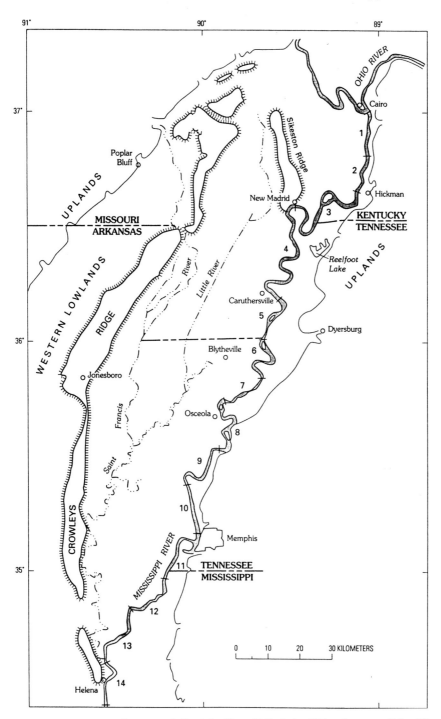

Figure 9.1 Index map of alluvial valley of Mississippi River between Cairo, Ill and Helena, Ark. Numbers indicate Mississippi River reaches that have been identified between Cairo and Helena.

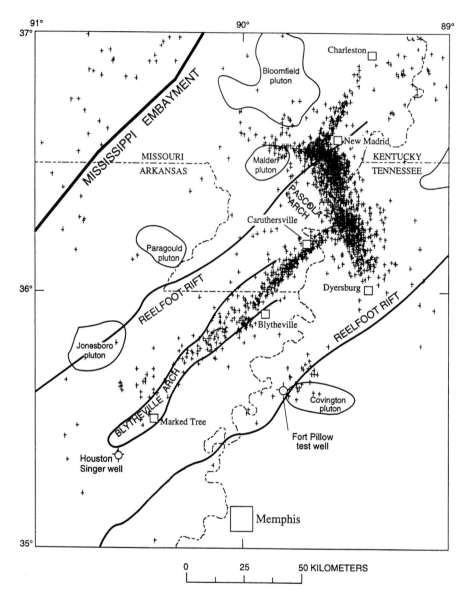

Figure 9.2 Regional tectonic features of the New Madrid seismic zone
showing plutons, the Blytheville arch, the Pascola arch, and epicenters
of microearthquakes (crosses) in the upper Mississippi Embayment
(modified from Luzietti and others, 1992).

been performed thus far has recently been compiled into maps for the area
between New Madrid and Memphis (Rhea and Wheeler, 1994). The major subsurface feature in the region is the Reelfoot Rift (Figure 9.2), a prominent
southwest-trending graben that extends from northeast of New Madrid,
Missouri, to southwest of the Tennessee–Mississippi border. Hildenbrand *et*

al. (1982) suggest that there are a series of smaller horsts and grabens within the rift floor that are a result of secondary faulting. Other structural features are the Pascola Arch and the Blytheville Arch (Figure 9.2), as well as several fault and tectonic zones.

The geomorphic investigation consisted primarily of the search for trends and anomalies on aerial photographs, topographic maps, and soil maps. Mississippi River maps dated 1765, 1820 to 1830, 1881 to 1893, and 1930 to 1932, and hydrographic surveys for 1880 and 1915 were used in the study. Data obtained from external sources were compiled and plotted by river mile or valley mile. River miles (RM) are measured along the course of the river, which includes all of the meanders. Valley miles (VM) are measured in a straight line that parallels the course of the river. Data provided by the U.S. Army Corps of Engineers was in English units, which is the common usage for navigation and flood-control purposes in the United States, and these units of feet and miles have been retained here.

Mississippi River anomalies

Changes of river pattern can be the first clue that active or past tectonics has affected the area of pattern change (Chapter 2). Of course, if more resistant rocks or coarser sediments appear in the channel on the crest of a structure, the river will respond to these non-alluvial materials, as it does to variations of floodplain sediment, such as clay plugs.

Based upon the evaluation of early maps and profiles of the Mississippi River between Cairo, Illinois and Old River, Louisiana, Schumm *et al.* (1994) divided the river into 24 reaches, of which 12, Reaches 3–14, are located within the area discussed here (Figure 9.1). These reach descriptions and divisions are based upon pre-cutoff program (pre-1930) channel morphology. This was necessary because the river was shortened 152 miles by 16 meander cutoffs, which were designed to reduce flood peaks. Individual reaches are characterized by different morphological characteristics and behavior, including channel-pattern differences, magnitude of pattern change through time (Figure 9.3), and changes of projected-profile slope, which is the water-surface elevation plotted against valley distance (Figure 9.4). For a discussion of projected profiles, see Chapter 4.

Near New Madrid, Missouri, the Mississippi River channel-dimension data indicate that locations of intense seismic activity during 1812 (see Chapter 5), as well as recent seismicity in the cross-rift seismic trend, are associated with reaches of anomalous channel morphology (Figure 9.3, 9.4) that are expressed by shallow depths and greater widths and width-to-depth ratios (Figure 9.5).

The abrupt steepening of the projected profile at the boundary between

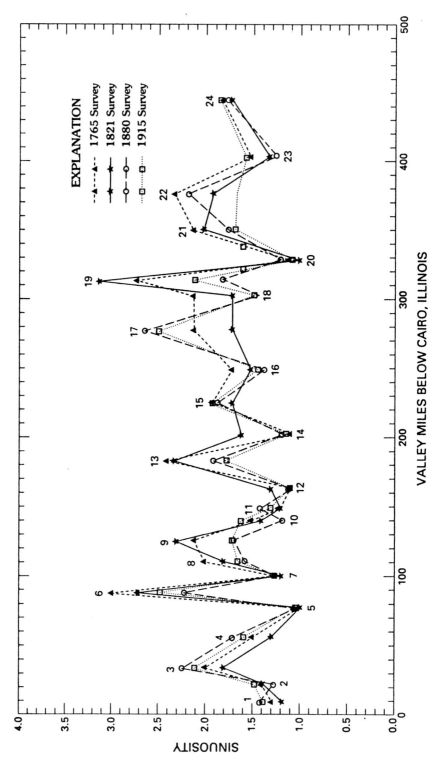

Figure 9.3 Sinuosity of 24 Mississippi River reaches (from Schumm *et al.*, 1994).

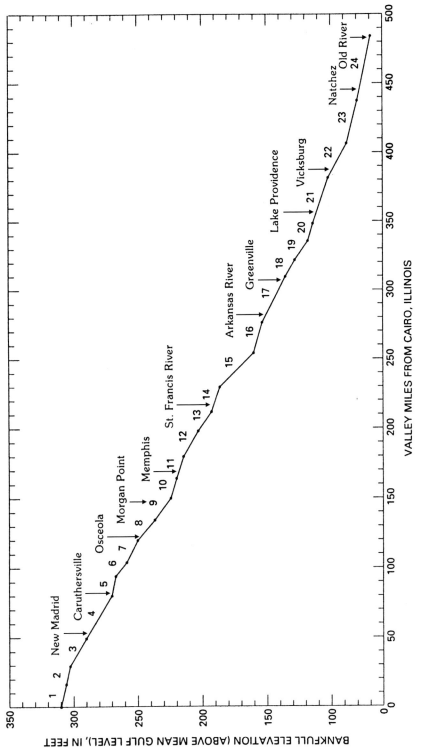

Figure 9.4 Projected-channel profile, consisting of 1880 bankfull-water-surface elevation plotted against valley miles (from Schumm *et al.*, 1994). Reaches 3 to 7 were included in this study.

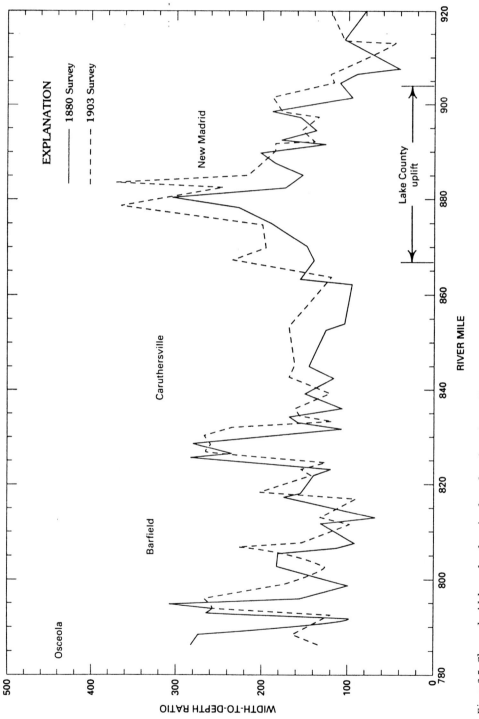

Figure 9.5 Channel width-to-depth ratio plotted against river mile for 1880 and 1903 surveys.

Reach 5 and Reach 6 (RM 826; Figure 9.4) and the compensating increase of sinuosity (Figure 9.3) and channel width-to-depth ratio (Figure 9.5), marks a dramatic change between Reaches 5, 6, and 7. Reach 6 is laterally active and sinuous relative to Reaches 5 and 7. The reach has an anomalously steep projected profile and a locally steepened water-surface profile at RM 820. A fault crossing has been proposed within the reach at RM 819 (Heyl and McKeown, 1978), and the behavior of the river here appears to confirm it.

Similar changes of the projected profile and stream pattern occur at the Reach 7 and Reach 8 boundary (Figures 9.3, 9.4) just north of Osceola, Arkansas (RM 791). The significant changes of Mississippi River character suggest that an external influence has modified the channel. No faulting has been detected, but the river encounters Tertiary-age sediments in the Chickasaw Bluffs, and this "bedrock" may be the controlling factor in this reach.

Another anomalously significant reach is Reach 9, which has historically been active and highly sinuous (Figure 9.3). The projected profile in this reach is fairly steep (Figure 9.4). The presence of inferred faults (RM 752 and 780, and the southeastern Reelfoot Rift margin) is coincident with the active and highly sinuous channel (Figure 9.3) and the steep projected profile of Reach 9 is suggestive of tectonic influences.

Farther downstream, Reaches 11 and 12 are characterized by an unusually low sinuosity (Figures 9.1, 9.3). Reaches 11, 12, and 13 trend southwest across the valley and Reach 13 represents the downstream end of a convexity in the water-surface profile that extends between Reaches 11 and 13 (Figures 9.4, 9.6). The trend of Reach 12 lies in close proximity to the strike of the Big Creek fault zone (Fisk, 1944). Both water-surface profiles and projected channel profiles indicate that deformation along Reaches 11, 12, and 13 is consistent with that of the Big Creek fault zone. As the river flows southwest onto the zone of uplift, as in Reach 11, the profile flattens and slopes are reduced by backwater effects resulting in aggradation. Reach 13 is on the downstream side of the uplifted area (Figures 9.4, 9.6). The river turns south in this reach, and it is characterized by historically high sinuosity (Figure 9.3).

The Mississippi River appears to be responding to either past or ongoing deformation of the valley floor. River morphology suggests that the New Madrid Seismic Zone, the Big Creek fault zone, the southeast rift margin, and other smaller inferred faults have and are affecting Mississippi River morphology and behavior. A fuller discussion of the Mississippi River reaches is presented by Schumm *et al.* (1994).

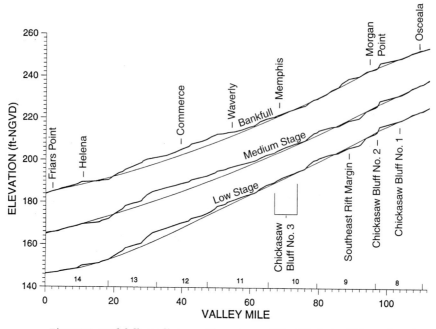

Figure 9.6 Bankfull, medium, and low stages of the Mississippi River plotted against valley mile. Reference line represents a hypothetical smooth profile.

Anomalies of the Western Alluvial Valley

If the Mississippi River responds to tectonically altered floodplain slopes, then smaller rivers on the adjacent floodplain should also respond in a similar fashion – this was demonstrated in Chapter 4 for several streams crossing the Monroe uplift in Louisiana and the Wiggins uplift in Mississippi. The major Western Lowlands (Figure 9.1) streams all show some form of disruption or anomalous behavior where they cross or pass close to major tectonic features. Anomalous course changes along linear trends not directly related to the major tectonic structures suggests subsidiary faulting or fracturing.

Black River

The Black River (Figure 9.7a) leaves the western bluff line at Poplar Bluff, Missouri. It flows southeast for 3 or 4 miles and then turns southwest to form a broad curve to the east. This pattern is mirrored by the St. Francis River, which is east of Black River (Figure 9.7a). Their arcuate courses are striking, but of more interest is the flow history of the northern half of Dan River (Figure 9.7b). Although Dan River is a continuous channel on Figure 9.7b, it is segmented by a drainage divide extending west from Big Island School. Because Dan River is an abandoned channel of the Black River, the flow should be to the south, however, topographic contours and tributary junctions indi-

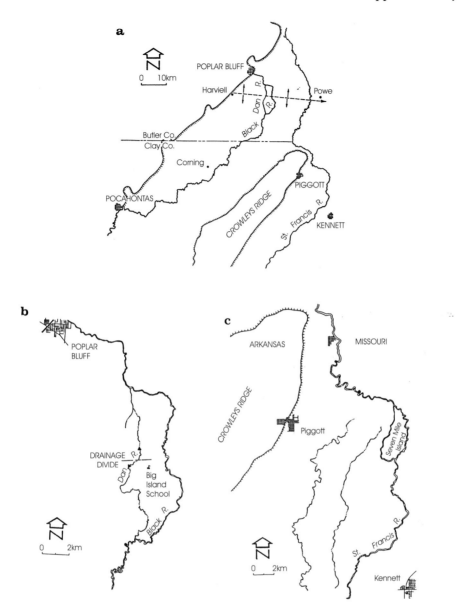

Figure 9.7 Maps showing anomalous river courses: (a) Black and St. Francis Rivers, dashed line shows location of inferred uplift; (b) Black and Dan Rivers, note that tributary to Dan River suggests flow to north; (c) present and abandoned courses of St. Francis River.

cate that part of the river flow is to the north (Harviell, Missouri quadrangle). The Dan, Black, and St. Francis Rivers' anomalous patterns suggest an upwarping along an east–west axis north of Big Island School (Figure 9.7b) and perhaps between Harviell and Powe, Missouri (Figure 9.7a).

Although the Black River's general trend is southwest in Figure 9.7a, its course is composed of a series of essentially north–south and east–west reaches. This angular pattern (rectangular drainage) is very similar to that of streams with courses controlled by faults or joints. Thus, the Black River contains river-pattern anomalies that suggest structural control, but the origin of the apparent control is unknown.

Little River anomalies

The Little River flows generally to the south past New Madrid and Lilbourn, Missouri, and turns abruptly west near Lonestar, Missouri (Figure 9.8). Its meander belt, as surveyed in 1870, was about 1 to 1½ miles wide. However, to the south, the meander belt narrows, and near the New Madrid–Pemiscot County line, the Little River makes a 6-mile deflection to the west past Wardell, Missouri, where the meander belt is only ¼ mile wide. It then turns abruptly south for 6 miles. East of Bragg City, Missouri, the meander belt widens abruptly to 2½ miles. These changes of direction and meander-belt width suggest that the river may be affected by the Pascola arch between Lonestar and Pascola (Figure 9.8). However, more recent investigations (Odum et al., 1998) reveal that a series of three faults have created the Little River anomalies.

St. Francis River

The St. Francis River (Figure 9.9) leaves the western bluff northeast of Poplar Bluff, Missouri and as noted above, it follows an arcuate course similar to the Black River to the south until it cuts southeastward through Crowleys Ridge (Figure 9.7a) at a point that coincides with the axis of the Pascola arch, as projected from the Caruthersville bend. After passing through Crowleys Ridge, it turns to the southeast and then back to the southwest (Figure 9.7a). Old courses of the St. Francis River southeast of Piggott, Arkansas, show that the river has progressively migrated to the east (Figure 9.7c). The bifurcation of the modern channel east of Piggott to form Seven Mile Island suggests that this eastward migration is continuing. Such avulsive changes can be a natural aspect of river behavior, but it can also reflect a subtle tilting of the floodplain to the east, which could be caused by uplift in the Ozark Mountains (McKeown et al., 1988).

Regional trends

The Mississippi River can be divided into reaches (Figure 9.1) that reveal the effect of the Lake County uplift (Reaches 3 and 4) and an inferred fault that crosses the river near the Reach 5 and 6 boundary (Figure 9.10), as mapped by Heyl and McKeown (1978). The Caruthersville bend appears to be

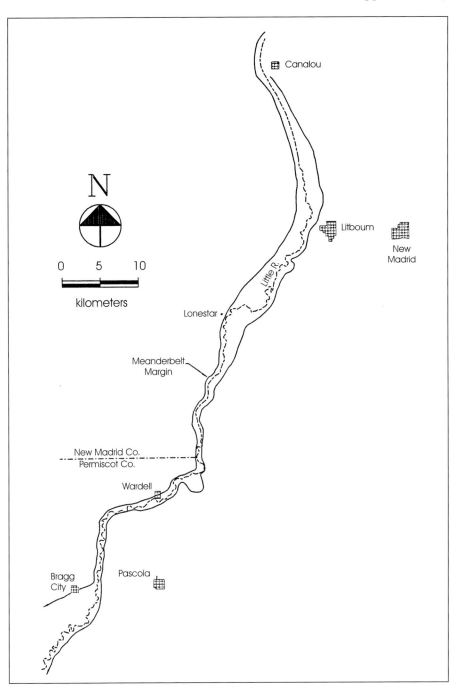

Figure 9.8 Map showing course of Little River and meander-belt width in the 1870s.

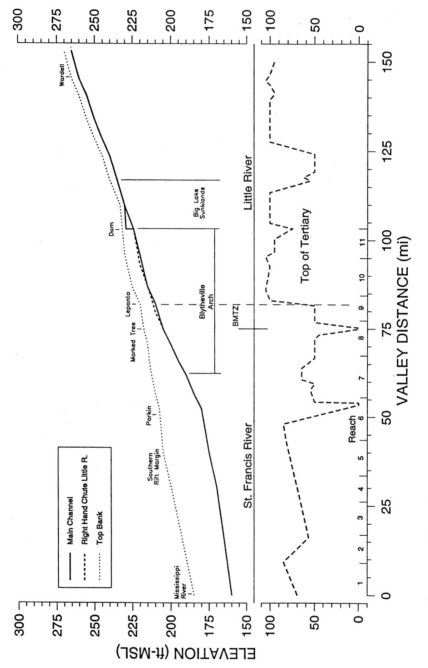

Figure 9.9 Profiles of the St. Francis and Little Rivers and the top of the Tertiary surface (from Saucier, 1964).

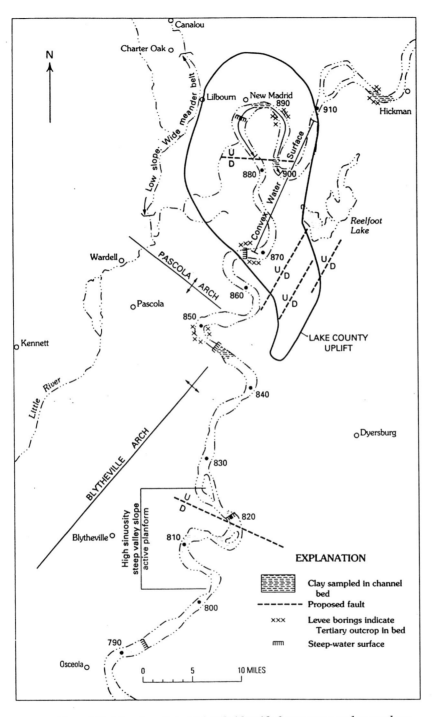

Figure 9.10 Map showing previously identified structures and anomalous geomorphic conditions along Mississippi River. Locations of geologic structures are from Russ (1982), O'Connell *et al.* (1982) and Heyl and McKeown (1978).

affected by Tertiary-age sediments that overlie the Pascola arch and Blytheville arch (Figure 9.10).

The Black River, northeast of Pocahontas, Arkansas, follows a peculiar angular course that suggests fracture control (Figure 9.7a). Other numerous anomalous abrupt changes in direction of the Black, St. Francis, and Little Rivers may be related to subsurface faulting or warping, as well as to differential compaction over structures or the effects of ground-water withdrawal. Nevertheless, many of these anomalies are more than local, and when considered on a larger scale, they suggest major structural trends (Figure 9.11). For example, between Piggott, Arkansas, and Pascola, Missouri, the St. Francis and Little Rivers flow in eastward arcs, having left abandoned channels to the west. This may be evidence of local uplift and eastward tilting. The Caruthersville bend is southeast of these features. This Mississippi River bend is anomalous as shown by its long-term stability as an extremely tight bend. The persistence of the bend is apparently due to the exposure of resistant sediments in the channel perimeter, which may be the result of uplift on the northwest-trending Pascola arch or northeast-trending Blythville arch.

Reaches along the St. Francis and Little Rivers show anomalous behavior that is coincident with structural features and is suggestive of tectonic influences. For example, the channel profile convexity and inflection of the top bank profile for both rivers are coincident with the Blytheville Arch (Figure 9.9), which, in turn, suggests that the structure has been recently active. Farther north, the St. Francis River and Black River flow in broad eastward arcs, suggesting regional eastward tilting, and the Little River displays a similar arc near Lilbourn, Missouri. Topographic plots of the Little River floodplain in this vicinity reveal local topographic depressions, which appear unrelated to depositional features and may reflect differential warping on an east–west trend. These anomalies are collectively shown as an east–west trend 2 in Figure 9.11. The Barfield bends of Reach 6 display anomalous geomorphic characteristics and are delineated as trend 3 in Figure 9.11. Within this reach, the slope of the projected profile is steep, the river channel is highly sinuous and historic planform maps depict extremely active channel migration and bend development. A fault has been mapped at this location (Heyl and McKeown, 1978); however, subsequent studies have not verified its existence. Trend 1, involving the St. Francis River watergap in Crowleys Ridge, and the anomalous Caruthersville bend with trends 2 and 3 (Figure 9.11) appear to represent structural controls on geomorphic features, that are normal to the major northeast–southwest structures associated with the Reelfoot rift.

The results of these investigations reveal numerous Mississippi River (Figures 9.3 to 9.10), Black River (Figure 9.7a), St. Francis River (Figure 9.7c),

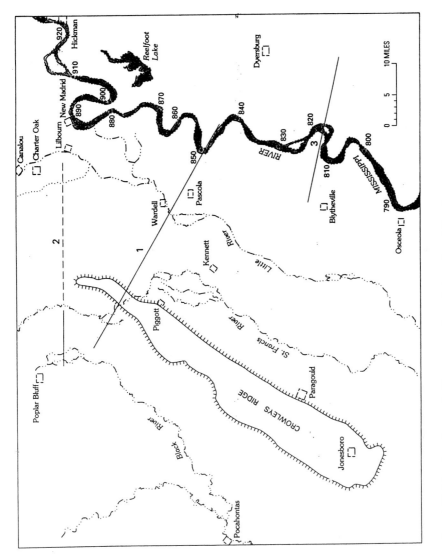

Figure 9.11 Map showing probable structural trends (numbered lines).

Little River (Figure 9.8), and other anomalies between Hickman, Kentucky and Osceola, Arkansas. The identification of these anomalies offers direction for scientists who are using subsurface geological and geophysical data to locate areas of tectonic deformation in the New Madrid seismic zone and other parts of the Mississippi River alluvial valley.

A zone of river anomalies

The 1886 Charleston, South Carolina earthquake was a large event (magnitude 6.6–6.9) on the coastal plain of the eastern U.S. However, as in the Mississippi valley, identification of tectonic features is difficult because a thick wedge of unconsolidated Cretaceous and Tertiary sediments blankets the deeper structures. According to Marple and Talwani (1993) and Marple (1994), surface expression of any active faults is quickly concealed by erosion, revegetation, and coastal processes. Nevertheless, these authors using geomorphic techniques have identified a 600 km long active fault system in the Coastal Plain of South Carolina, North Carolina, and Virginia (Marple and Talwani, 1993). They refer to this structure as the East Coast Fault System (ECFS), which coincides with a "zone of river anomalies" (ZRA, Figure 9.12), first identified as a 200 km long 10–15 km wide zone of anomalous river behavior in South Carolina (Marple and Talwani, 1993).

Marple and Talwani (1993) identified anomalies associated with channel incision, valley longitudinal profile, valley cross-section, channel sinuosity, anastomosing reaches, and course deflection (Figures 9.12 and 9.13). As further proof of the active nature of the ZRA and ECFS, five rivers to the south of the ECFS were studied and they do not show these anomalous characteristics. Marple and Talwani have identified what appears to be a major and very important structural feature using fluvial geomorphology. Available geologic and geophysical data support their conclusions.

Sacramento River

Between the Sierra Nevada and the Coast Ranges lies the synclinal trough of the Great Valley of California, the axis of which is off-center to the west. In this structure there are many major folds and faults that parallel the Sierra Nevada. The Sacramento River drains the northern half of this major structural trough entering it at Red Bluff and leaving it at San Francisco Bay.

Within the Great Valley there are petroleum and natural gas fields, which indicate that there are important structural features beneath the valley alluvium. If these structures have an influence on the alluvial surface they should

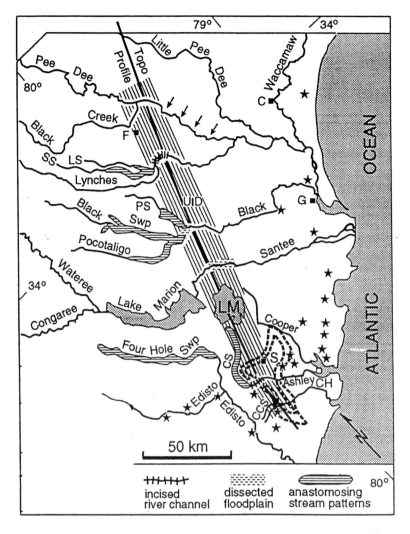

Figure 9.12 Rivers of South Carolina Coastal Plain with zone of river anomalies (ZRA, striped area), anastomosing stream patterns, pre-1886 sand blow sites (stars), isoseismals of the 1886 Charleston, South Carolina earthquake and dashed closed contours near Summerville (S), and bold line shows approximate axis of ZRA. The arrows along the north side of the Pee Dee River downstream of the ZRA denote that part of the river that is flowing against the southwest side of its valley. U/D denotes location of buried fault inferred on EXXON seismic reflection profile (unpublished data). C, Conway; CCS, Caw Caw Swamp; CH, Charleston; CS, Cypress Swamp; F, Florence; G, Georgetown; LM, Lake Moultrie; LS, Lake Swamp; PS, Pudding Swamp; S, Summerville; SS, Sparrow Swamp (from Marple and Talwani, 1993).

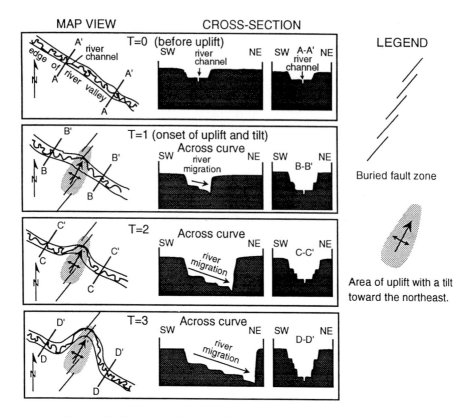

Figure 9.13 Hypothetical model of how the northward-convex curves in the river valleys along the ZRA (Figure 9.12) may have formed. Before uplift (T = 0) the river channel meandered across a relatively mature floodplain between relatively steep, symmetric valley walls. Uplift along a buried fault combined with a tilt toward the northeast causes the river to erode downward and locally toward the northeast side of its valley, subsequently becoming entrenched against its northeast valley wall (T = 1). Continued uplift and tilting toward the northeast results in a local curve in the river valley that has an asymmetric cross-valley shape with unpaired terraces or northward-sloping terrain on the southeast side of the curved valley (T = 2, T = 3). Cross-sections A–A′ to D–D′ show the vertical entrenchment of the river valley both upstream and downstream from the ZRA in response to sea level fluctuations (from Marple and Talwani, 1993).

also affect the morphology and behavior of the Sacramento River in the reach from Red Bluff to Colusa (Figure 9.14), which was studied in order to provide the Corps of Engineers with information on bank stability and flooding (Fischer, 1994).

Because of the active tectonic nature of California and especially the 1975 Oroville earthquake northeast of Colusa, it seems reasonable to assume that deformation is occurring at present. The possibility that this reach of the Sacramento River is affected by active deformation of the Great Valley can be

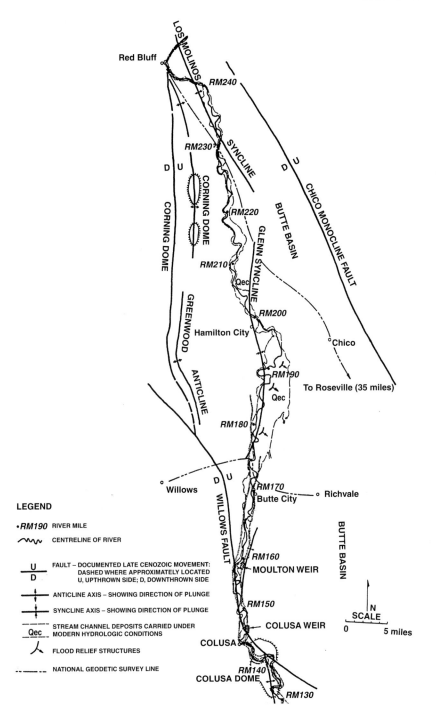

Figure 9.14 Structural map of Sacramento Valley from Red Bluff to Colusa (from Harwood and Helley, 1987) showing National Geodetic Survey lines and flood relief structures (from Fischer, 1994).

addressed through re-levelling data from the National Geodetic Survey (NGS). Levelling lines are available from Red Bluff to Roseville (Figures 9.14, 9.15) and from Willows to Richvale (Figure 9.16). The Red Bluff to Roseville line, which is located on the east side of the Sacramento River, was originally surveyed in 1919, and it was resurveyed in 1949 (30 years). However, it only provides information on the northernmost part of the river course. Figure 9.15 indicates that there has been subsidence, with the greatest amount to the south. The data also suggest that there is active subsidence where the survey crosses the northern tip of the Glenn syncline. The Willows to Richvale line runs from west to east across the river and the southern part of the Butte Basin (Figures 9.14, 9.16). It was surveyed originally in 1935, and it was resurveyed in 1949 (14 years). The data indicate that there has been deformation over the Willows fault, and there has been subsidence to the east of the river. This may explain why the Butte Basin to the east of the river has persisted as a naturally occurring flood overflow area.

Although major faults dominate the structural geology of the northern part of the Sacramento valley, it is the smaller anticlines and synclines that seem to control the course and perhaps even the present behavior of the Sacramento River (Harwood and Helley, 1987, p. 16). For example, the Sacramento River flows toward the southeast after leaving its canyon at Red Bluff. At about RM 245 it encounters the Los Molinos syncline, and it occupies the axis of the syncline at RM 240 (Figure 9.14). At this point, the width of the active channel and floodplain of the river widen significantly; they remain wide to about RM 233, where they narrow as the river turns south and leaves the syncline at about RM 230. The change of width of the active channel deposits is striking, and it is associated with the Los Molinos syncline. Similar changes of floodplain width occur as the river is affected by the Glenn syncline to the south (Figure 9.14).

In a geomorphic study of the Sacramento River, Schumm and Harvey (1985) identified 17 reaches that showed different river characteristics. The selection of the reaches was initially based only on channel pattern and floodplain width; nevertheless, marked valley-slope changes are associated with the boundaries of Reaches 8–17 (Figure 9.17), for which good-quality profile data were available (U.S. Congress, 1911). Reaches 10, 12, 14, and 16 are very gentle whereas Reaches 8, 9, 11, 13, 15, and 17 are steep. The gentle reaches are the least active. The steep reaches contain the greatest number of bends, evidence of cutoffs, and probably the greatest reduction of channel width following the construction of Shasta Dam upstream. The Sacramento River appears to be responding to changes of valley slope as did the experimental channels (Chapter 3).

One means of demonstrating the relative stability or lack of it in the

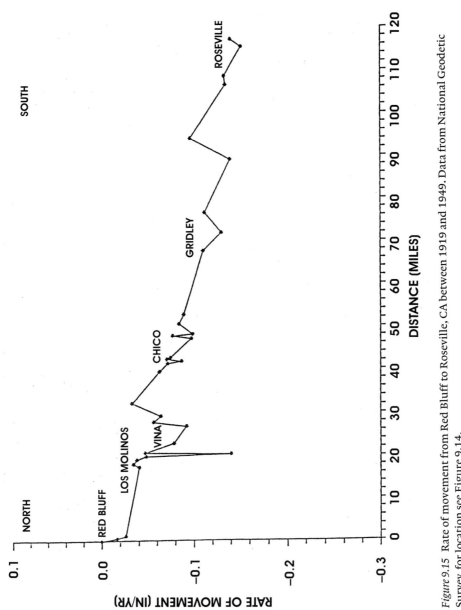

Figure 9.15 Rate of movement from Red Bluff to Roseville, CA between 1919 and 1949. Data from National Geodetic Survey, for location see Figure 9.14.

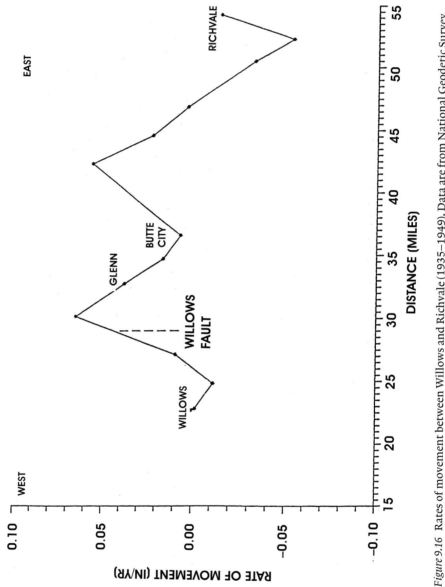

Figure 9.16 Rates of movement between Willows and Richvale (1935–1949). Data are from National Geodetic Survey, see Figure 9.14 for location.

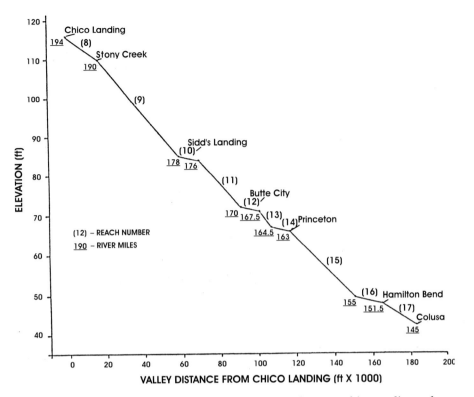

Figure 9.17 Valley profile along Sacramento River between Chico Landing and Colusa. Elevations were obtained from a longitudinal profile of low-water in 1909 and locations of reach boundaries were estimated on the Department of Water Resources Atlas (Schumm and Harvey, 1985).

reaches, is to plot the sinuosity for each reach, as measured for the 1908 and 1976 channel patterns (State of California, 1984). The relatively steep reaches show very marked changes in sinuosity during this period of approximately 70 years (Figure 9.18), whereas the gentle reaches (10, 12, 14, 16) have remained straight and stable.

Starting at Chico Landing (RM 194) the projected profile (Figure 9.17) can be compared with the occurrence of minor structural features (Figure 9.14). The projected profile shows a marked decrease of gradient at Sidds Landing (RM 178–176), where the river leaves the influence of the Glenn syncline (Figure 9.17). South of Butte City (RM 170), the width of the active channel deposits decrease, and the river appears to be confined between the Willows fault and a small unnamed anticline, and at RM 163 to RM 145 the river flows on the fault line of the Willows fault to Colusa (RM 145) where the river crosses the Colusa Dome. There is a decrease in gradient upstream (Reaches 16, 17) that may be associated with the upstream effect of the Colusa Dome.

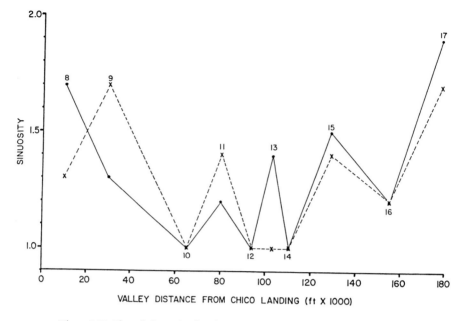

VALLEY DISTANCE FROM CHICO LANDING (ft X 1000)

Figure 9.18 Plot of sinuosity for river reaches between Chico Landing and Colusa. Data points are plotted in the middle of each reach. Solid line is 1908 sinuosity; dashed line is 1976 sinuosity.

The irregularities of the projected profile are probably the result of these structures. In summary, both the projected profile and the width of the active channel deposits appear to be affected by the minor structural features of the Sacramento Valley.

Observations of the Sacramento River floodplain support the conclusion of Gomez and Marron (1991) concerning the width of floodplain deposits. For example, the river flows south after leaving the Molinos syncline, with the Corning Domes to the west, until it enters the Glenn syncline at about RM 205 (Figure 9.14). The width of the active channel deposits widen, as the river enters the syncline, and then it narrows at RM 200 to RM 197 where the river crosses the axis of the syncline. The active channel deposits then widen at RM 197, as the river turns abruptly west and then south to follow the axis of the syncline to RM 173, where the active channel deposits narrow again.

In summary, Reaches 1, 2, 4, 6, 7, 8, 9, 10, and 13 are closely associated with components of the minor structures of the Sacramento River valley. Only Reaches 3, 5, and 11 have no apparent structural control. Structural control is possible and, in fact, it is likely in Reaches 12, 14, 15, 16, 17. In conclusion, association of river characteristics with the minor structures of the Great Valley indicates that the structures are still influencing river morphology and behavior in the study reach, and the NGS surveys indicate that at least some of these structures are presently active and significantly affecting the Sacramento River.

Nile River

The Nile River within Egypt is 1350 km long. For the first 300 km, the river is in a narrow valley until it reaches the First Cataract near Aswan (Figure 9.19). This portion of the Nile valley upstream of the High Aswan Dam (HAD) is submerged beneath Lake Nasser. North of Aswan, the valley widens to an average of 10 km between Aswan and Cairo (Said 1990), but it is about 20 km wide in places downstream of Qena (Figure 9.20). Bedrock controls the river at Aswan and perhaps at Silsila Gorge just to the north of Kom Ombo and, of course, bedrock controls the right (eastern) bank of the river at many locations in the valley.

The river hugs the eastern valley wall below the Qena bend (Figure 9.20) and Butzer (1976) concludes that during historic time most changes of river position have been to the east. Old maps and descriptions support this conclusion, as do abandoned natural levees to the west. Youssef (1968) concludes that the Nile Valley is controlled by faults; the orientation of major fracture patterns and most of the Nile valley trends are related to the Red Sea rifting patterns, and gravity surveys north of 27°N latitude have also identified numerous subsurface faults (Riad, 1977) that cross the valley at depth. Additional evidence for possible modern deformation is seismic activity that is concentrated in the Nile valley (Gergawi and El Khashab, 1968; Said, 1981).

In addition to faults, areas of uplift and subsidence may affect the valley and river. Of particular significance are the Assiut Basin north of El Manshah (Figure 9.20) and the probability that a major structure crosses the river near El Baalyana (P.G. Vigano and W.M. Meshref, personal communication). The cliffs on the western side of the valley near Assiut, 260 km downstream of Qena, are lower than those on the east (Said, 1981), which suggests that the position of the river in its valley (Figure 9.20) is caused by tilting to the east.

An unusual aspect of the River Nile is that both gradient and median grain size of the bed sediment increase downstream. These downstream changes are anomalous because the reverse is true for alluvial rivers unless there is a major impact of a large tributary, which is not the case in Egypt, or the river is affected by bedrock or tectonics. Petroleum geologists believe a major structure crosses the river near El Manshah (P.G. Vigano and W.M. Meshref, personal communication). Examination of the water surface slope and the projected profile (Figure 9.21) reveals a significant increase of both valley and channel slope near this location. The slope of the valley increases by about 25 percent from 7.6 cm/km to 9.8 cm/km.

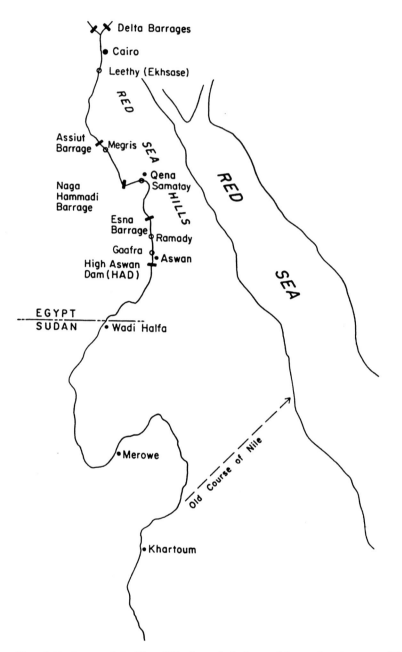

Figure 9.19 Course of the River Nile through Sudan and Egypt showing a possible former course and irrigation barrages between Aswan and the Delta.

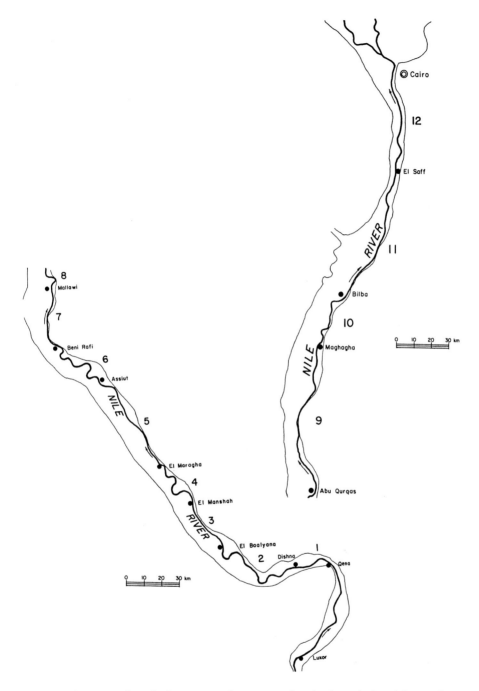

Figure 9.20 The Nile downstream from Luxor showing boundaries of the Reaches
1–12. Numbers identify geomorphic reaches (from Schumm and Galay, 1994).

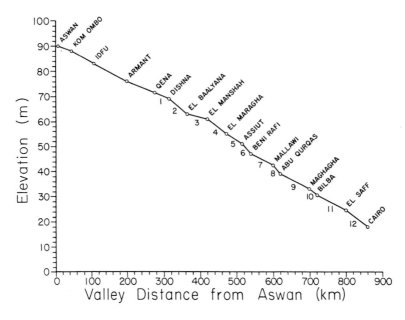

Figure 9.21 Valley profile from Aswan to Cairo. Numbers identify geomorphic reaches. Prepared by Mr. Hussein El Kubtan of the Nile Research Institute.

In addition to a division of the river into two long reaches at El Manshah, other reaches can be identified by changes of channel pattern and slope. The River Nile is remarkably straight, but there are sinuous reaches, although the change in all cases occurs for only relatively short distances, and it may be slight. The projected profile (Figure 9.21) reveals that there are 12 reaches between Qena and Cairo, and six of these have higher than average valley slope. Four of these reaches are clearly more sinuous than the rest of the river (Figure 9.22).

The increased projected profile slope in each case may be related to geologic structure as well as to wadi sediment contributions (Reach 2). The 25 percent increase of valley slope and channel gradient at El Manshah is probably related to the downstream Assiut structural basin. This increase of valley slope should have caused a general increase of sinuosity downstream, but it did not. The explanation for this apparent anomaly is given by Figure 9.22, which is based upon the relation between sinuosity and valley slope that was developed experimentally and confirmed for other rivers (Figure 2.3). This plot of sinuosity and projected profile slope shows that the River Nile channel becomes sinuous only at the steepest valley slopes, about 12 or 13 cm/km. The increased valley slope at El Manshah is not sufficient to produce a continuously meandering river downstream of that location.

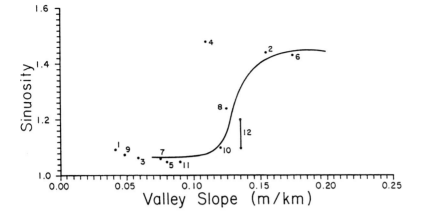

Figure 9.22 Plot of sinuosity (ratio of channel length to valley length) and valley slope. The range of sinuosity between 1800 and 1990 is shown for Reach 12. Numbers identify geomorphic reaches (from Schumm and Galay, 1994).

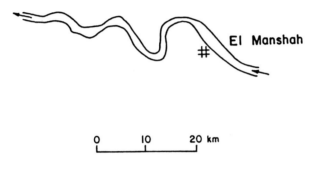

Figure 9.23 River pattern near El Manshah.

However, the El Manshah reach (4) has a sinuosity that is apparently far too high for the valley slope (Figure 9.23). The channel pattern in this reach is one of decreasing sinuosity and meander amplitude in a downstream direction (Figure 9.23). This pattern is similar to that obtained during experimental studies of river patterns (Schumm *et al.*, 1987). When a straight channel was forced to meander under these conditions, the sinuosity and meander amplitude decreased in the same fashion as in the El Manshah reach. Although the El Manshah reach is steeper that the upstream and downstream reaches, its high sinuosity is probably the effect of a perturbation of the channel, probably by faulting at El Manshah. Therefore, the threshold slope that produces meandering in the Nile valley is probably in excess of 12 cm/km (Figure 9.22). The logical conclusion is that Reach 4 should straighten with time.

Channel adjustment to a steeper slope in the Nile valley also appears to

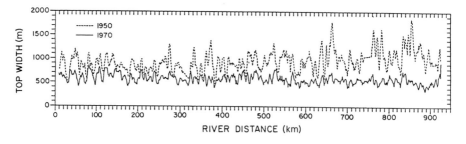

Figure 9.24 Variability of width between Aswan and Cairo for pre-HAD (1950) and post-HAD (1970) conditions. Data have been smoothed by plotting the 5 km moving mean (from Evans and Attia 1990).

occur by an increase of width (Figure 9.24). The pre-HAD width and variability of width increase from around Dishna (RK 314) to El Wasta (RK 844). A glance at Figure 9.22 reveals that an increase of valley slope from 10 to 15 cm/km, a 50 percent increase, results in an increase of sinuosity from about 1.05 to 1.4, which causes only a 33 percent decrease of gradient. The remainder of the adjustment to the increase of valley slope must occur in width, depth, and roughness. The most unresponsive of the reaches in terms of sinuosity was Reach 10, and the mean width increased from 1010 m in Reach 9 to 1330 m in Reach 10. It is likely that the general increase of width below Dishna is a reflection of the increased slope of the valley floor.

In summary, the Nile is a relatively straight and stable river. Morphologic variability can be attributed to changes of valley slope, which cause adjustments of sinuosity and width. The projected profile changes reflect wadi contributions and geologic structures (Figure 9.21).

Hydraulic analysis, Alabama River

The Alabama River basin includes approximately 27 750 square miles and extends from eastern Tennessee and northwestern Georgia diagonally across Alabama. The Alabama River and Tombigbee River join about 45 miles upstream of Mobile Bay to form the Mobile River (Figure 9.25).

Geologic structures near the Alabama River include the South Carlton salt dome and the regional peripheral fault trend (Figure 9.25). A series of 867 surveyed cross-sections of the river channel was made available by the U.S. Army Corps of Engineers. The surveyed cross-sections are located approximately every 0.1 river mile, from Claiborne Lock and Dam downstream to the confluence with the Tombigbee River. Hydraulic characteristics of the river were determined through the use of the HEC-2 hydrologic model, which performs step backwater computations at a given discharge.

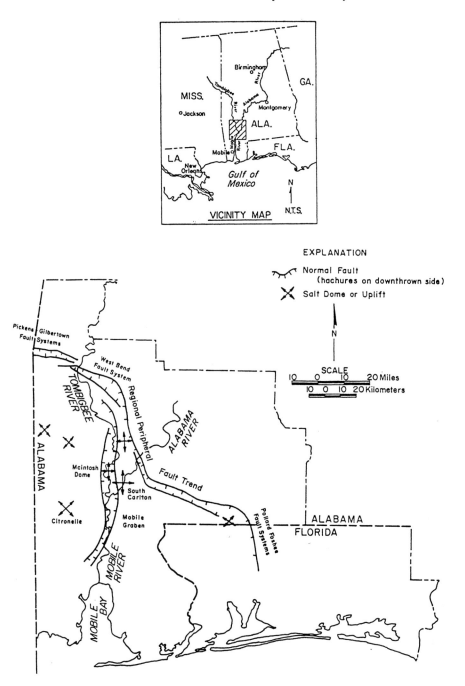

Figure 9.25 Major structural features of southwest Alabama.

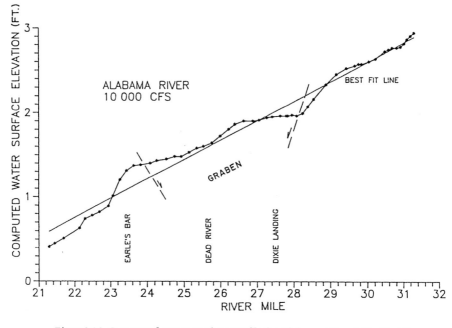

Figure 9.26 Computed water-surface profile for Alabama River Mile 21–32. Discharge is 10 000 cfs (from Harvey and Schumm, 1994).

Every published geologic map of southwestern Alabama shows the peripheral fault trend, but they show it crossing beneath the Alabama River at different locations. Several authors (e.g., Scott, 1972) depict the faults dying out before reaching the river, whereas numerous other sources show the fault zone beneath the river downstream of Choctaw Bluff (Martin, 1978). The results of HEC-2 modeling of the river provide an opportunity to locate this structure.

The computed water-surface elevations are plotted against river mile so that they represent channel slope (Figure 9.26). Water-surface elevations decrease downstream, but two significant deviations from this downstream trend occur at about RM 24 and RM 29. The relative directions of these deviations are opposite; at RM 29, the water surface drops below the trend, and at RM 24 it rises above the general trend. These two locations are approximately three valley miles apart, which is the approximate width of the peripheral fault trend. It is, therefore, suggested that Figure 9.26 shows the downstream flow of the Alabama River into and out of the fault zone. Just upstream of Dixie Landing, the river crosses a normal fault onto a downthrown block, and at Earle's Bar it leaves the downthrown block.

In summary, the HEC-2 modeling has located the fault trend where it crosses the Alabama River. In regions where subsurface geology is unknown, hydrologic modeling results can aid in the delineation of structural features along a river course. This evidence has important economic implications in terms of locating domal structures that are potential traps for liquid hydrocarbons. The delineation of fault zones is possible as well. As geologic maps do not show the location of faults beneath alluvial valleys, determining the location of active faults by hydrologic modeling may provide a valuable tool for geologic mapping.

Discussion

The geomorphic evaluation has identified several anomalous surface features along the valley floor in the New Madrid seismic zone, some of which can be directly linked to mapped structures in the region. Of the numerous other features, some may be the result of influences from non-structural controls such as sediment variability, and tributary influences; however, many apparently result from structural controls. The many geomorphic anomalies probably reflect a densely fractured suballuvial surface (Figure 9.27), as in East African rift valleys (Dart and Swolfs, 1994). Movement on these fractures probably varies in space and time. Although most activity will occur in the seismically active zones, the possibility of movement elsewhere is high. The rivers and streams of the Mississippi Embayment, the zone of river anomalies in South Carolina, the Central Valley of California, the Nile valley, and the Alabama River valley, are flowing on alluvium and, therefore, should be isolated from structural patterns in the underlying Cenozoic rocks. However, the anomalous patterns identified herein suggest that the suballuvial structures have a surface expression. The densely fractured suballuvial surface is apparently readily adjustable to increased or decreased stress. For example, the greatest number of microseisms occur during rising and falling stages of Mississippi River floods (McGinnis, 1963). It is known that the 1811–1812 earthquakes occurred during a high flood period. It is possible that the loading and subsequent unloading of the valley alluvium in this fashion causes movement along existing fractures that are propagated to the surface and significantly affect river patterns and behavior (Costain et al., 1987). Deformation may be slow and aseismic or may be rapid and episodic.

The numerous river anomalies in the Mississippi Embayment on the Atlantic Coastal Plain and in the Sacramento River Valley, provide an excellent example of the application of geomorphology to the detection of major

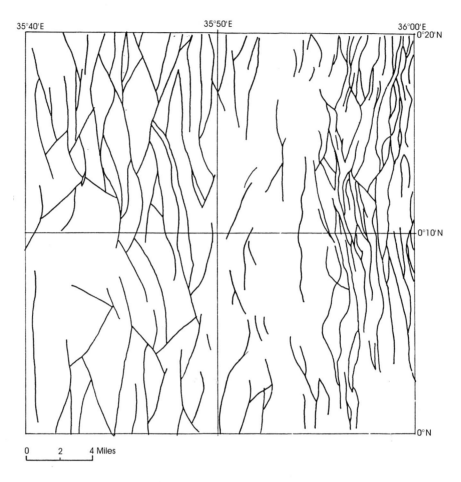

Figure 9.27 Fault patterns on the floor of an East African rift valley.

structural features. The use of hydraulics to locate faults in the Alabama River valley suggests that geomorphology (Chapter 4), sedimentology, hydrology, and hydraulics (Chapter 6) can all be employed in the search for syntectonic controls on river behavior, which, in turn, provides information on tectonically active structures.

10

Applications

The discussions in the previous chapters are convincing that active tectonics modifies alluvial river behavior, hydrology, hydraulics, and sedimentology. As a result, there can be little doubt that the influence of active tectonics on alluvial rivers is important for human activities. By restricting our discussions to alluvial rivers, we are dealing with a most sensitive landform and, therefore, river response will be rapid, in contrast to rivers constrained by bedrock or more resistant terrace alluvium. The field examples and experiments indicate how sensitive these rivers can be to slight changes of valley-floor slope. Channel morphology, therefore, appears to provide a means of detecting active tectonics, subsidence due to groundwater and petroleum withdrawal, and differential compaction of sediments over bedrock highs. When the changes of valley-floor slope continue, as a result of active tectonics, river adjustment must be continuous or episodic. In either case the effects should be clearly displayed by geomorphic and hydrologic evidence of deformation as follows:

1. Drainage-network change (Chapters 1, 2, 3, 7, 9)
2. Channel-pattern change (Chapters 1, 2, 3, 4, 7, 9)
3. Channel-dimension change (Chapters 4, 5, 6, 9)
4. Water-surface profile variability (Chapters 4, 5, 9)
5. Channel instability (aggradation, degradation) or variability (Chapters 6, 9)
6. Overbank-flow frequency change or variability (Chapters 4, 6, 9)
7. Lake morphology and change (Chapter 2)
8. Progressive lateral shift and avulsion (Chapter 7)

When anomalous conditions such as those listed above are recognized, historical information and a sequence of maps and aerial photographs can be used to track the changes and the rate of change. Geophysical information on subsurface features will be especially valuable in determining if the evidence truly relates to active tectonics or to human or other causes. In addition, an understanding of how deformation affects rivers and fluvial sediments will aid in the interpretation of ancient fluvial deposits that can be attributed to syntectonic deposition (Chapter 8). Therefore, we hope that the preceding discussions will have practical applications to the present (siting of structures, river engineering, navigation, flood frequency) and to the past (sedimentology, stratigraphy).

Siting of structures

Until it is recognized that slow aseismic deformation can cause progressive degradation, aggradation, pattern change, lateral shift and even avulsion, the solution to many river stability problems may not be effective. For example, continued uplift and tilting can significantly influence overbank flooding and bank erosion that can threaten highways, bridges, and other structures. For example, Dumont (1994) describes the loss of a power plant in Peru as a result of persistent lateral shift of the Ucayali River. Recognition that such a tendency will persist as a result of lateral tilt could result in selection of a different site for the structure. The siting of hazardous waste disposal or storage sites also requires that the potential for tectonic instability be recognized. Not only local areas can be affected, but even ancient civilizations. For example, canals are usually constructed to carry relatively clear water on gentle slopes. A slight warping can seriously affect the efficiency of a canal (Lees, 1955; Leary *et al.*, 1981) and indeed Moseley (1983) and Adams (1965) have proposed that the decline of agricultural societies in Peru and Iraq were the result of tectonics, which disrupted irrigation systems. In addition, avulsion of the Indus River undoubtedly led to abandonment of Harappan cities on the Indus plain. An awareness of the potential for deformation and its subsequent impacts should be part of the perspective of planners and engineers as they select sites for urban, and industrial uses.

River stabilization and navigation

If the net result of active tectonics is unstable river reaches, that can be characterized by incision, deposition, bank erosion, meander cutoffs, or development of meanders, braided or anastomosing patterns, then it is very important for the river engineer to understand this phenomenon. For example, river and harbor navigation can be affected by aggradation and the development of bars. Bank erosion, incision, and pattern change can impact on riparian use and cause loss of valuable structures (bridges, loading docks).

If a serious navigation problem exists because of deposition of sediment in the thalweg, as a result of downstream uplift, dredging of the deposited sediment is only a palliative measure. Instead, when the deformation is recognized, the channel on the axis of uplift should be dredged. That is, the cause of upstream deposition should be removed in order that sediment transport is not obstructed. Also, identification of river reaches affected by deformation will permit the concentration of resources in those areas.

In the future, further attempts to identify active areas within the Lower Mississippi valley are desirable. Both the Pearl River and the Red River flow over recognized uplift features and their associated fault systems. It is certain that these two rivers, plus the Ouachita River, and portions of the lower Mississippi River are influenced significantly by active tectonics. Planning for navigation and flood control projects should consider the effects of active tectonic movement.

Hydrology

The studies of high-water marks along the Gulf Coast streams and the frequency of flooding in the Central Valley of California reveal that areas subjected to greater than normal flooding are located on the upstream limb and axis of an uplift. Perhaps little can be done to improve the situation, but at least these areas should not be sites of development. Also, changes in the frequency of overbank flooding through a zone of active tectonics will change the position of ordinary high water, which is usually a legal boundary, and this may lead to confusion and litigation concerning the location of the river bank, which may be a property or state boundary.

Geology

The numerous river anomalies in the Mississippi Embayment on the Atlantic Coastal Plain and in the Sacramento River Valley, provide an excellent example of the application of geomorphology to the detection of major structural features (Chapter 9). The use of hydraulics to locate faults in the Alabama River valley suggests that geomorphology (Chapters 4, 9), sedimentology, hydrology, and hydraulics (Chapter 6) can all be employed in the search for syntectonic controls on river behavior, which, in turn, provides information on tectonically active structures, and their impacts on diverse types of rivers. Obviously, a recognition and understanding of how active tectonics influences river behavior will be of value in the interpretation of ancient fluvial deposits (Chapter 8).

For example, subtle facies variations owing to tectonic influences on fluvial sedimentation can determine the locations of concentrations of economically important petroleum reservoirs and coal seams.

Petroleum

Leeder (1993) points out the relevance to the petroleum industry of tectonic controls on sand body distribution in asymmetric basins. Because of short transport distances, rapid subaerial deposition and abundant debris flows, sand deposited in transverse systems (e.g., basin-marginal alluvial fans and braid plains) tend to be texturally immature, and make poor reservoirs (Leeder, 1993). Reservoir quality of such strata may be improved however if processes like fault and fold migration force these deposits to be uplifted and reworked, thereby providing second-cycle deposits. In contrast, the reservoir potential of axial-channel belts is considerably higher, because perennial-stream processes tend to promote better sorting. The position of these axial channel belts tend to be controlled by the extent of transverse alluvial deposits and the presence of any intrabasinal antithetic or synthetic faults (Leeder, 1993).

The fluvial Upper Mississippian Kekiktuk Formation in the Endicott field (Alaska North Slope) is a producing reservoir in an asymmetric graben bound on the northeast down-dropped margin by the Mikkelson Bay and Tigvariak faults. The sand-rich channel-belt deposits are concentrated near the bounding Mikkelson Bay and Tigvariak faults of this tilted basin. Melvin (1993) noted this is probably due to shifting of axial rivers toward the axis of greatest subsidence (Bridge and Leeder, 1979).

Controls on paleovalley position by tectonic uplift have considerably influenced hydrocarbon distribution in the U.S. Western Interior. Weimer *et al.* (1986) discuss the role of these tectonically controlled valleys on production from the Muddy Sandstone in the Wattenburg field of Denver Basin. The Muddy Sandstone reflects a system of valley fills (Horsetooth member) incised into shoreface sandstone (Ft. Collins member). The position of the paleovalleys is controlled by uplift on contemporary structures, which diverted incising rivers into adjacent tectonically low areas.

Similar tectonic controls on valley position and hydrocarbon production occur throughout Muddy strata of the northern Western Interior Basin. Dolson *et al.* (1991) note that production regionally is greatest in tectonically low areas where the valley-fill sandstone is thickest, although interfluve areas where petroleum is trapped beneath overlying unconformities remains important. Similar structural controls on valley position have been instrumental in the development of reservoirs from Lower Pennsylvanian Morrow channels of southeast Colorado and Western Kansas as well (Scott and Horne, 1989; Thomas, 1993).

Coal

Fielding (1987) suggests that subsidence rate is an overriding factor (compared to coarse sediment supply and depositional environment) in preservation of peat. Where subsidence is too slow, peat may not have sufficient accommodation space to accumulate. If subsidence is too fast, excessive seam splitting or coarse fill will probably result. A moderate amount of subsidence is thus optimal for preservation of quality coal. Fielding's (1987) points are supported by subsequent studies. Titheridge (1993), for instance, saw such relationships in the Brunner Coal Measures in New Zealand. These strata were deposited in a tilting half-graben, by meandering streams, and in coastal lakes. Coals parallel the northwest-trending Mangatini and west-trending Millerton faults, which are located along the basin margin. Areas where subsidence exceeded peat growth are characterized by seam splits. Seam splits up to 50 m thick are most common near the Mangatini Fault where subsidence was greatest. Seam splits appear to represent areas where peat failed to keep pace with sedimentation, and channels migrated into these areas of lower topography (Titheridge, 1993).

Localized uplift within the depositional basin may also have a significant impact on peat accumulation. Johnson (1989) discussed the impact of a system of NE/SW trending folds on coal beds of the Triassic Yan'an Formation in the Ordos Basin of China. Effects of folds on coal deposition are manifest in four ways: (1) coal beds merge toward anticline axes and split toward syncline axes; (2) individual coal beds thin over anticlines; (3) split thickness decreases over anticlines; and (4) ash content increases over anticlines.

Intrabasinal folds may not only control peat thickness, but peat distribution as well. Wise *et al.* (1991) explain five continuous coal beds in the Pennsylvanian Allegheny Formation of the Appalachian Basin using a diversion model. They argue that contiguous swamps of 40 000 km² are needed to make some of these coal beds and that fold salients diverted major trunk streams around the swamps. These areas were thus free of major clastic input, and were excellent sites for extensive coal swamps. Sinking of the foreland near thrusts provided the necessary subsidence. As salients breached, clastic influx increased, and the channel belt facies that cap these coals were deposited. They cited modern examples of such diversions by fold salients in the Zagros and Himalayan mountains.

Belt (1993) similarly argues that large fold salients (e.g., Sequatche anti-cline) in the Appalachian thrust front formed barriers that diverted trunk streams exiting the fold and thrust belt around vast areas. He argued that thick extensive coal in middle and early Late Pennsylvanian deposits of the Appalachian foredeep resulted from such swamps. When salients were later breached, clastics were introduced, making for thin coals with abundant part-ings (e.g., Kittanming and Freeport coal zones, Pennsylvania). He noted that the coally zones had channel-belt spacing on the order of 75 km, whereas over-lying coarser deposits with less coal had channel belts 10–30 km apart.

A final word

We hope that the material presented here will be useful in several areas of endeavor; nevertheless, we realize that geomorphic evidence may lead an investigator into erroneous conclusions. For example, Stein and King (1984) used four lines of evidence to demonstrate uplift during the 1983 Coalinga, California earthquake. One piece of evidence was that Los Gatos Creek where it crossed the Coalinga anticline changed from slightly sinuous to straight. Atwater *et al.* (1990) disregarded this evidence because it was the reverse of what was expected from Ouchi's (1985) results. Their more detailed study contradicted the earlier work, but Figure 2.3 indicates that with increased valley slope, a meandering or a channel transitional between mean-dering and braided can straighten and braid. Therefore, the apparent contra-diction as noted by Atwater *et al.* (1990), can still be used to support the conclusion that uplift has occurred. Clearly, changes of river pattern are easy to detect on maps, aerial photographs, or in the field, but these changes are only suggestive and additional evidence must be obtained to support any con-clusion about tectonic activity. For example, the presence of differential aggra-dation and degradation in association with geologic structures is not certain evidence of active tectonics. The Canning River, which rises in the Brooks Range of northern Alaska and flows across the Arctic coastal plain, demon-strates a repeated change of channel pattern from braided to meandering and from erosion to deposition, apparently coincident with geologic structures that have been active in the late Quaternary. However, the presence of three end moraines marking the latest glacial advances affect the profile of the Canning River and are the true reason for the change in erosion and deposi-tion, as well as pattern change. Therefore, a problem often encountered is in determining which channel changes are the result of tectonics and which are the result of other factors.

References

Adams, J. (1980). Active tilting of the United States midcontinent: geodetic and geomorphic evidence. *Geology*, **8**, 442–6.

Adams, R. M. (1965). *Land behind Baghdad*. University of Chicago Press, Chicago, IL, 187 pp.

Alexander, J. and Leeder, M. R. (1987). Active tectonic control on alluvial architecture. In Ethridge, F. G., Flores, R. M., and Harvey, M. D. (eds.) *Recent developments in fluvial sedimentology*. Society of Economic Paleontologists and Mineralogists Special Pub. No. 39, pp. 243–52.

Alexander, J., Bridge, J. S., Leeder, M. R., Collier, R. E. L., and Gawthorpe, R. L. (1994). Holocene meander-belt evolution in an active extensional basin, southwestern Montana. *J. Sed. Res.*, **B64**, 542–59.

Allen, J. R. L. (1965). A review of the origin and characteristics of recent alluvial sediments. *Sedimentology*, **5**, 89–191.

Allen, J. R. L. (1978). Studies in fluvial sedimentation: an exploratory quantitative model for the architecture of avulsion-controlled alluvial suites. *Sed. Geol.*, **21**, 129–47.

Allenby, R. J. (1988). Origin of rectangular and aligned lakes in the Beni Basin of Bolivia. *Tectonophysics*, **145**, 1–20.

Ambraseys, N. N. (1978). Studies in historical seismicity and tectonics. In *The environmental history of the Near and Middle East*. Academic Press, New York, pp. 185–210.

Anadón, P., Cabrera, L., Colombo, F., Marzo, M., and Riba, O. (1986). *Syntectonic intraformational unconformities in alluvial fan deposits, eastern Ebro Basin margins (NE Spain)*. International Association of Sedimentologists Special Publ. 8, pp. 259–71.

Andrews, J. T. (1991). Late Quaternary glacial isostatic recovery of North America, Greenland and Iceland: a neotectonics perspective. In Slemmons, B. D. (ed.) *The decade of North America geology*. Geological Society of America, pp. 473–86.

Angelier, J., Dumont, J. F., Poisson, A., Simsek, S. and Karamanderesi, I. H. (1981). Neotectonic evolution of the Southwestern part of Turkey. *Tectonophysics*, **75**, 1–9.

Assumpçào, M. (1992). The regional intraplate stress field in South America. *J. Geophys. Res.*, **97**(B8), 11889–903.

Assumpçào, M. and Araujo, M. (1993). Effect of the Altiplano-Puna plateau, South America, on the regional intraplate stresses. *Tectonophysics*, **221**, 475–93.

Assumpçào, M. and Suarez, G. (1988). Source mechanisms of moderate size earthquakes and stress orientation in mid-plate South America. *Geophys. J.* **92**, 253–67.

Atwater, B. F., Trumm, D. A., Tinsley, J. C., Stein, R. S., Tucker, A. B., Donahue, D. J., Jull, A. J. T., and Payen, L. A. (1990). *Alluvial plains and earthquake recurrence at the Coalinga anticline*. U. S. Geol. Survey Prof. Paper 1487, pp. 273–98.

Autin, W. J., Burns, S. F., Miller, B. J., Saucier, R. T., and Snead, J. I. (1991). Quaternary geology of the Lower Mississippi valley. In Morrison, R. B. (ed.) *The geology of North America, v. K-2: Quaternary nonglacial geology conterminous U.S.*, Geol. Soc. Am., pp. 547–81.

Bachman, G. O. and Mehnert, H. H. (1978). New K-Ar dates and the late Pliocene to Holocene geomorphic history of the central Rio Grande region, New Mexico. *Geol. Soc. Am. Bull.*, **89**, 283–92.

Bagnold, R. A. (1980). An empirical correlation of bedload transport in flumes and natural rivers. *Proc. Roy. Soc. Lond.*, **A372**, 453–73.

Bailey, G., King, G., and Sturdy, D. (1993). Active tectonics and land-use strategies: a Palaeolithic example from northwest Greece. *Antiquity*, **67**, 292–312.

Baker, V. R. (1978). Adjustment of fluvial system to climate and source terrain in tropical and subtropical environments. In Miall, D. (ed.) *Fluvial sedimentology*. Canadian Society of Petroleum Geology, Memoir 5, pp. 211–30.

Barnes, V. E. (1968). Palestine sheet: Scale 1:2500,000. *Geologic Atlas of Texas*. Univ. of Texas, Bureau of Economic Geol., Austin, TX.

Barrell, J. (1917). Rhythms and the measurement of geologic time. *Geol. Soc. Am. Bull.*, **28**, 745–904.

Barry, J. M. (1997). *Rising tide: the great Mississippi flood of 1927 and how it changed America*. Simon and Schuster, New York.

Baumgardner, R. W. (1987). *Landsat-based lineament analyses, east Texas Basin and Sabine Uplift area*. Univ. Texas, Bureau of Economic Geol., Report of Investigations 167, 26 pp.

Basu, A., Young, S. W., Suttner, L. J., James, W. C., and Mack, G. H. (1975). Reevaluation of the use of undulatory extinction and polycrystallinity in detrital quartz for provenance interpretation. *J. Sed. Petrol.*, **45**, 873–82.

Beaumont, C. (1981). Foreland Basins. *Roy. Astron. Soc. Geophys. J.* **65**, 291–329.

Beck, R. A. (1985). Syntectonic sedimentation adjacent to Laramide basement thrusts, Rocky Mountain foreland: timing of deformation. M. S. Thesis, Iowa State Univ., Ames, IA, 89 pp.

Beck, R. A. and Vondra, C. F. (1985). Syntectonic sedimentation and Laramide basement thrusting. In *Abstracts, International Symposium on Foreland Basins*. International Assoc. Sedimentologists, Fribourg, Switzerland, p. 36.

Beck, R. A., Vondra, C. F., Filkins, J., and Olander, J. (1988). Syntectonic sedimentation and Laramide basement thrusting: timing of deformation. In Schmidt, C. J. and Perry, W. J. (eds.), *Interaction of the Cordilleran thrust belt and Rocky Mountain foreland*. Geological Society of America, Memoir 171.

Belt, E. S. (1993). *Tectonically induced clastic sediment diversion and the origin of thick, widespread coal beds (Appalachian and Williston basins, USA)*. International Assoc. Sedimentologists, Special Pub. 20, pp. 337–97.

Bendefy, L., Dohnalik, J., and Mike, K. (1967). Nouvelles méthodes de l'étude génétique des cours d'eau. *Int. Assoc. Sci. Hydrol.*, Pub. 75, 64–72.

Benito, G., Perez-Gonzalez, A., Gutierrez, F., and Machado, M. J. (1997). River response to Quaternary subsidence due to evaporite solution (Gallego River, Ebrobasin, Spain). *Geomorphology*, **22**, 243–63.

Bergerat, F. (1988). Evolution des mecanismes d'extension dans le bassin pannonien. *Geodinamica Acta, Paris,* **2**(2), 89–98.

Bergerat, F. (1989). From pull-apart to the rifting process: the formation of the Pannonian Basin. *Tectonophysics,* **157**, 271–80.

Blair, T.C. (1987). Tectonic and hydrologic controls on cyclic alluvial fan, fluvial, and lacustrine rift-basin sedimentation, Jurassic-lowermost Cretaceous Todos Santos Formation, Chiapas, Mexico. *J. Sed. Petrol.,* **57**(5), 845–62.

Blair, T.C. and Bilodeau, W.L. (1988). Development of tectonic cyclothems in rift, pull-apart, and foreland basins: sedimentary response to episodic tectonism. *Geology,* **16**, 517–20.

Blair, T.C. and McPherson, J.G. (1994). Historical adjustments by Walker River to lake-level fall over a tectonically tilted half-graben floor, Walker Lake Basin, Nevada. *Sed. Geol.,* **92**, 7–16.

Blake, D.H. and Ollier, C.D. (1970). Geomorphological evidence of Quaternary tectonics in southwestern Papua. *Rev. Geomorph. Dynam.,* **19**, 28–32.

Blakey, R.C. and Gubatosa, R. (1984). Controls of sandstone body geometry and architecture in the Chinle Formation (Upper Triassic), Colorado Plateau. *Sed. Geol.,* **38**, 51–86.

Blum, M.D. (1993). Genesis and architecture of incised valley fill sequences: a late Quaternary example from the Colorado River, Gulf Coastal Plain of Texas. In Weimer, P. and Posamentier, H.W. (eds.) *Siliclastic sequence stratigraphy.* American Association of Petroleum Geology, Memoir 58, pp. 259–83.

Blum, M.D. and Valastro, S., Jr (1994). Late Quaternary sedimentation, lower Colorado River, Gulf Coastal Plain of Texas. *Geol. Soc. Am.,* **106**, 1002–16.

Blum, M.D., Toomey, R.S., III, and Valastro, S., Jr (1994). Fluvial response to Late Quaternary climate and environmental change, Edwards Plateau, Texas. *Palaeogeog., Palaeoclimatol., Palaeoecol.,* **108**, 1–21.

Borsi, Z. and Flegyhzi, E. (1983). Evolution of the network of water courses in the northeastern part of the Great Hungarian Plain from the end of the Pleistocene to our days. *Quatern. Studies Poland,* **4**, 115–24.

Bowler, J.M. and Harford, L.B. (1966). Quaternary tectonics and the evolution of the Riverine Plain near Echuca, Victoria. *Geol. Soc. Austral.,* **13**, 339–54.

Boyce, J.K. (1990). Birth of a megaproject: political economy of flood control in Bangladesh. *Environ. Man.,* **14**, 419–28.

Boyd, K.F. and Schumm, S.A. (1995). *Geomorphic evidence of deformation in the northern part of the New Madrid seismic zone.* U.S. Geol. Surv. Prof. Paper 1538–R, 35 pp.

Bradley, W.C. (1963). Large-scale exfoliation in massive sandstones of the Colorado Plateau. *Geol. Soc. Am. Bull.,* **74**, 519–28.

Braga, G. and Gervasoni, S. (1989). Evolution of the Po River: an example of the application of historic maps. In Petts, G.E. (ed.) *Historical change of large alluvial rivers.* Wiley, New York, pp. 113–26.

Bridge, J.S. (1984). Large-scale facies sequences in alluvial overbank environments. *J. Sed. Petrol.,* **54**(2), 0583–8.

Bridge, J.S. (1985). Paleochannel patterns inferred from alluvial deposits: a critical evaluation. *J. Sed. Petrol.,* **55**(4), 0579–89.

Bridge, J.S. and Diemer, J.A. (1983). Quantitative interpretation of an evolving ancient river system. *Sedimentology,* **30**, 599–623.

Bridge, J. S. and Leeder, M. R. (1979). A simulation model of alluvial stratigraphy. *Sedimentology*, **26**, 617–44.

Bridge, J. S. and Mackey, S. D. (1993). A revised alluvial stratigraphy model. In Marzo, M. and Puigdefábregas, C. (eds.) *Alluvial sedimentation*. International Association of Sedimentologists Special Pub. 17, pp. 319–36.

Brown, R. D., Jr and Wallace, R. E. (1968). *Current and historic fault movement along the San Andreas fault between Pacines and Camp Dix, CA*. Stanford Univ. Pub., Geological Science, vol. 11, pp. 22–41.

Bryant, M., Falk, P., and Paola, C. (1995). Experimental study of avulsion frequency and rate of deposition. *Geology*, **23**(4), 365–8.

Bull, W. B. (1984). Tectonic geomorphology. *J. Geol. Ed.*, **32**, 310–24.

Bull, W. B. and Kneupfer, P. L. K. (1987). Adjustments of the Charwell River, New Zealand, to uplift and climatic change. *Geomorphology*, **1**, 15–32.

Burbank, D. W., Beck, R. A., Raynolds, R. G. H., Hobbs, R., and Tahirkheli, R. A. K. (1988). Thrusting and gravel progradation in foreland basins: a test of post-thrusting gravel dispersal. *Geology*, **16**, 1143–6.

Burnett, A. W. (1982). Alluvial stream response to neotectonics in the lower Mississippi valley. Unpublished M. S. thesis, Colorado State Univ., Fort Collins, CO, p. 160.

Burnett, A. W. and Schumm, S. A. (1983). Alluvial river response to neotectonic deformation in Louisiana and Mississippi. *Science*, **222**, 49–50.

Butzer, K. W. (1976). *Early hydraulic civilization in Egypt*. University of Chicago Press, Chicago, IL.

Cant, D. J. and Abrahamson, B. (1996). Regional distribution and internal stratigraphy of the Lower Mannville. *Can. Petrol. Geol. Bull.*, **44**(3), 508–29.

Cant, D. J. and Stockmal, G. S. (1993). Some controls on sedimentary sequences in foreland basins: examples from the Alberta Basin. In Frostick, L. E. (ed.) *Tectonic controls and signatures in sedimentary successions*. International Association of Sedimentologists Special Publication 20, pp. 49–65.

Carpenter, C. H., Robinson, G. G., and Bjorklund, L. J. (1967). *Groundwater conditions and geologic reconnaissance of the upper Seiver River basin*. U.S. Geological Survey, Water-Supply Paper 1836, 91 pp.

Cartier, K., and Alt, D. (1982). The bewildering Bitterroot, the river that won't behave. *Montana Mag.*, **12**, 52–3.

CERESIS (1985). *SISRA Project, preliminary neotectonic map of South America*. CERESIS, Lima.

Clark, J. A., Farrell, W. E., and Peltier, W. R. (1978). Global changes in postglacial sea level: a numerical calculation. *Quat. Res.*, **9**, 265–87.

Clemente, P. and Pérez-Arlucea, M. (1993). Depositional architecture of the Cuerda Del Pozo Formation, Lower Cretaceous of the extensional Cameros Basin, North-central Spain: *J. Sed. Petrol.*, **63**(3), p. 437–52.

Collins, E. W., Dix, O. R., and Hobday, D. K. (1981). *Oakwood Salt Dome, east Texas, surface geology and drainage analysis*. University of Texas, Bureau of Economic Geology Circular, 81–6, 23 p.

Costain, J. K., Bollinger, G. A., and Speer, J. A. (1987). Hydroseismicity – A hypothesis for the role of water in the generation of intraplate seismicity. *Geology*, **15**, 618–21.

Cox, R. T. (1994). Analysis of drainage-basin symmetry as a rapid technique to identify areas of possible Quaternary tilt-block tectonics: an example from the Mississippi embayment. *Geol. Soc. Am. Bull.*, **106**, 571–81.

Crittenden, M. D. (1963). *New data on the isostatic deformation of Lake Bonneville*: U.S. Geol. Survey Prof. Paper 454–E, 31 pp.

Csontos, L., Nagymarosy, A., Horvath, F. and Kovac, M. (1992). Tertiary evolution of the intra-carpathian area: a model. *Tectonophysics*, **208**, 221–41.

Dart, R. L. and Swolfs, H. S. (1994). Structural style of the Reelfoot Rift. *Geol. Soc. Am. Abst. Prog.*, **26**(1), 5.

Davis, W. M. (1899). The geographical cycle. *Geog. J.*, **14**, 481–504.

Dawson, M. (1988). Sediment size variation in a braided reach of the Sunwapata River, Alberta, Canada. *Earth Surface Processes and Landforms*, **13**, 599–618.

DeBlieux, C. (1949). Photogeology of Gulf Coast exploration. *Am. Assoc. Petrol. Geol. Bull.*, **33**, 1251–9.

DeBlieux, C. (1962). Photogeology in Louisiana coastal marsh and swamp. *Gulf Coast Assoc. Geol. Soc. Trans.*, **12**, 231–41.

DeBlieux, C. W. and Shephard, G. F. (1951). Photogeology study in Kent County, TX. *Oil Gas J.*, **50**, 86, 88, 98–100.

DeCelles, P. G. (1986). Sedimentation in a tectonically partitioned nonmarine foreland basin: the Lower Cretaceous Kootenai Formation, southwestern Montana. *Geol. Soc. Am. Bull.*, **97**, 911–31.

Deffontaines, B. and Chorowicz, J. (1991). Principles of drainage basin analysis from multisource data: application to the structural analysis of the Zaire Basin. *Tectonophysics*, **194**, 237–63.

Dickinson, W. R. and Suczek, C. A. (1979). Plate tectonics and sandstone composition. *Am. Assoc. Petrol. Geol. Bull.*, **94**, 222–35.

Dingman, S. L. (1984). *Fluvial hydrology*. W. H. Freeman and Company, New York, 383 pp.

Dolson, J. C., Muller, M. J., Evetts, M. J., and Stein, J. A. (1991). Regional paleotopographic trends and production, Muddy Sandstone (Lower Cretaceous), central and northern Rocky Mountains. *Am. Assoc. Petrol. Geol. Bull.*, **75**(3), 409–35.

Doornkamp, J. C. (1986). Geomorphological approaches to the study of neotectonics. *Geol. Soc. Lond. J.*, **143**, 335–42.

Doornkamp, J. and Temple, B. (1966). Surface drainage and tectonic instability in part of southern Uganda. *Geog. J.*, **132**, 238–52.

Doser, D. I. (1985). Source parameters and faulting processes of the 1959 Hebgen Lake, Montana, earthquake. *J. Geophys. Res. B*, **90**, 4537–55.

Dumont, J. F. (1992). Rasgos morfoestructurales de la llanura Amazonica del Peru: efecto de la Neotectonica sobre los cambios fluviales y la delimitacion de las provincias morfologicas. *Bull. de l'Institut Francais d'Etudes Andines*, **21**(3), 801–33.

Dumont, J. F. (1993). Lake patterns as related to neotectonics in subsiding basins: the example of the Ucamara Depression, Peru. *Tectonophysics*, **222**, 69–78.

Dumont, J. F. (1994). Neotectonics and rivers of the Amazon headwaters. In Schumm, S. A. and Winkley, B. R. (eds.) *The variability of large alluvial rivers*. American Society of Civil Engineers Press, New York, pp. 103–14.

Dumont, J. F. (1996). Neotectonics of the Subandes–Brazilian craton boundary using geomorphological data: the Marañón and Beni basins. *Tectonophysics*, **257**, 137–151.

Dumont, J. F. and Fournier, M. (1994). Geodynamic environment of Quaternary morphostructures of the subandean foreland basins of Peru and Bolivia. *Quat. Int.*, **21**, 129–42.

Dumont, J. F. and Garcia, F. (1991). Active subsidence controlled by basement structures in the Maranon Basin of northeastern Peru. *Land Subsidence, Proceedings of the Fourth International Symposium on Land Subsidence*, International Assoc. Hydrological Sciences, no. 200, pp. 343–50.

Dumont, J. F. and Guyot, J. L. (1993). Ria lac: ou, pourquoi? *Proceedings of the Third International Conference on Geomorphology*, 23–29 August, 1993, Hamilton, Canada.

Dumont, J. F. and Hanagarth, W. (1993). River shifting and tectonics in the Beni Basin (Bolivia), *Proceedings of the Third International Conference on Geomorphology*, 23–29 August, 1993, Hamilton, Canada.

Dumont, J. F., Lamotte, S., and Fournier, M. (1988). Neotectonica del Arco de Iquitos (Jenaro Herrera, Peru). *Boletin de la Sociedad Geologica del Peru*, **77**, 7–17.

Dumont, J. F., Deza, E., and Garcia, F. (1991a). Morphostructural provinces and neotectonics in Amazonian lowlands of Peru. *S. Am. Earth Sci.*, **4**(4), 287–95.

Dumont, J. F., Herail, G., and Guyot, J. L. (1991). Subsidencia, inestablidad fluvial y reparticion de los placeres distales de oro, el caso del Rio Beni (Bolivia). *Symposium International sur les Gisements Alluviaux d'Or*, 3–5 June, 1991, LaPaz, Bolivia.

Dumont, J. F., Garcia, F., and Fournier, M. (1992). Registros de cambios climaticos por los depositos y las morfologias fluviales en la Amazonia Occidental. In Ortlied, L. and Machare, J. (eds.) *Proceedings of the International Symposium on Paleo Enso Records, Lima, Abstracts*, pp. 87–92.

Dunne, T. (1988). Geomorphologic contributions to flood control planning. In Baker, V. R., Kochel, R. C., and Patton, P. C. (eds.), *Flood geomorphology*. Wiley and Sons, NY, pp. 421–38.

Dury, G. H. (1970). General theory of meandering valleys and underfit streams. In Dury, G. H. (ed.) *Rivers and River Terraces*, pp. 264–75, MacMillan, London.

Dykstra, S. (1988). The effects of tectonic deformation on laboratory drainage basins. Unpublished report, Colorado State Univ.

Ego, F., Sebrier, M., Lavenu, A., Yepes, H., and Egues, A. (1996). Quaternary state of stress in the Northern Andes and the restraining bend model for the Ecuadorian Andes. *Tectonophysics*, **259**, 101–16.

Elliot, T. (1974). Interdistributary bay sequences and their genesis. *Sedimentology*, **21**, 611–22.

Ellis, M. and Merritts, D. (1994). *Tectonics and topography*: a collection reprinted from *J. Geophys. Res. Solid Earth*, **99**, B6, B7, B10. American Geophysical Union, Washington, D.C.

Espejo, I. S. and López-Gamundí, O. R. (1994). Source versus depositional controls on sandstone composition in a foreland basin: the El Imperial Formation (Middle Carboniferous–Lower Permian), San Rafael Basin, Western Argentina. *J. Sed. Res.*, **A64**(1), 8–16.

Ethridge, F. G. and Schumm, S. A. (1978). Reconstructing paleochannel morphologic and flow characteristics: methodology, limitations and assessment. In Miall, A. D. (ed.) *Fluvial sedimentology*. Canadian Soc. Petroleum Geology Memoir 5, pp. 703–21.

Evans, B. J. and Attia, K. (1990). Changes to River Nile channel properties after High Aswan Dam. *National seminar on physical response of the River Nile to interventions, proceedings*, Cairo, unpaginated.

Fairchild, H. L. (1932). Earth rotation and river erosion. *Science*, **76**, 423–7.

Farrand, W. R. (1962). Post glacial uplift in North America. *Am. J. Sci.*, **260**, 181–99.

Ferguson, H. F. (1974). Geologic observations and geotechnical effects of valley stress release in the Allegheny Plateaus: Preprint, American Society of Civil Engineers, *Water Research Engineers Meeting, Los Angeles, CA*, 31 pp.

Fielding, C. R. (1987). Coal depositional models for deltaic and alluvial plain sequences. *Geology*, **15**, 661–4.

Fielding, C. R., Falkner, A. J., and Scott, S. G. (1993). Fluvial response to foreland basin overfilling; the Late Permian Rangal Coal Measures in the Bowen Basin, Queensland, Australia. *Sed. Geol.*, **85**, 475–97.

Fischer, K. J. (1994). Fluvial geomorphology and flood strategies: Sacramento River, California. In Schumm, S. A. and Winkley, B. R (eds.) *The variability of large alluvial rivers*. American Society of Civil Engineers Press, New York, pp. 115–38.

Fisk, H. N. (1944). *Geologic investigation of the alluvial valley of the lower Mississippi River*. Vicksburg, MS, U.S. Army Corps of Engineers, 78 pp.

Fisk, H. N. (1947). *Fine-grained alluvial deposits and their effect on Mississippi River activity*. Mississippi River Commission, Vicksburg, MS.

Flam, L. (1993). Fluvial geomorphology of the lower Indus basin (Sindh, Pakistan) and the Indus civilization. In Shroder, J. F., Jr (ed.), *Himalaya to the sea*: Routledge, London, pp. 265–87.

Flemings, P. B. and Jordan, T. E. (1989). A synthetic stratigraphic model of foreland basin development. *J. Geophys. Res.*, **94**, 3851–66.

Folk, R. L. (1974). *Petrology of sedimentary rocks*. Hemphill, Austin, TX, 182 pp.

Freeland, G. L. and Dietz, R. S. (1971). Plate tectonic evaluation of the Caribbean – Gulf of Mexico region. *Nature*, **232**, 20–3.

Fuller, M. L. (1912). The New Madrid earthquake. *U.S. Geol. Surv. Bull.*, **494**, 119 pp.

García-Gil, S. (1993). The fluvial architecture of the upper Buntsandstein in the Iberian Basin, central Spain. *Sedimentology*, **40**, 125–43.

Gardner, T. W. (1975). *The history of part of the Colorado River and its tributaries: an experimental study*. Four Corners. Geological Society Guidebook, Canyonlands, pp. 87–95.

Gaudemer, Y., Tapponnier, P., and Turcotte, D. (1989). River offsets across active strike-slip faults. *Ann. Tectonicae*, **III**, 55–76.

Gawthorpe, R. L. and Colella, A. (1990). Tectonic control on coarse-grained delta depositional systems in rift basins. In Colella, A. and Prior, D. B. (eds.) *Recent and ancient nonmarine depositional environments: models for exploration*. Society Economic Paleontologists and Mineralogists Special Pub. 31, pp. 127–55.

Gay, S. P. Jr. (1994). The Basement fault block pattern: Its importance in petroleum exploration, and its delineation with residual aeromagnetic techniques: *10th International Conference on Basement Tectonics, Proceedings*, pp. 159–207.

Gergawi, A. and El Kashab, H. M. A. (1968). Seismicity of the UAR. *Helwan Observ. Bull.*, **76**, 1–27.

Germanoski, D. and Schumm, S. A. (1993). Change in braided river morphology resulting from aggradation and degradation. *J. Geol.*, **101**, 451–66.

Gilbert, G. K., 1877). *Report on the geology of the Henry Mountains*. U.S. Geographical and Geological Survey of the Rocky Mountain Region, 160 pp.

Gilchrist, A. R. and Summerfield, M. A. (1991). Denudation, isostasy and landscape evolutions. *Earth Surf. Proc. Landforms*, **16**, 555–62.

Gomez, B. and Marron, D.C. (1991). Neotectonic effects on sinuosity and channel migration, Belle Fourche River, western South Dakota. *Earth Surface Processes and Landforms*, **16**, 227–35.

Gordon, I., and Heller, P.L. (1993). Evaluating major controls on basinal stratigraphy, Pine Valley, Nevada: implications for syntectonic deposition. *Geol. Soc. Am. Bull.*, **105**, 47–55.

Gourou, P. (1949). *Observcoes geograficas na Amazonia*. Rev. Brasil. Geogr.

Greb, S.F. and Chesnut, D.R. (1996). Lower and lower Middle Pennsylvanian fluvial to estuarine deposition, central Appalachian basin: effects of eustasy, tectonics, and climate. *Geol. Soc. Am. Bull.*, **108**(3), 303–17.

Gregory, D.I. and Schumm, S.A. (1987). The effect of active tectonics on alluvial river morphology. In Richards, K.S. (ed.) *River channels: environment and process*. Blackwells, London, pp. 41–68.

Grunthal, G. and Stromeyer, D. (1992). The recent crustal stress field in Central Europe: trejectories and finite element modeling. *J. Geophys. Res.*, **97**(B8), 11805–20.

Guccione, M.J. and Hehr, L.H. (1991). Origin of the "Sunklands" in the New Madrid seismic zone: tectonic or alluvial drowning. *63rd Annual meeting Eastern Section Seismological Society of America Program and Abstracts*, p. 64.

Guccione, M.J. and Van Arsdale, R.B. (1994). *Origin and age of the St. Francis sunklands using drainage patterns and sedimentology*, National Earthquake Hazard Reduction Program, Summaries of Technical Reports Volume XXXV, U.S. Geol. Survey Open-File Report No. 94–176, **1**, 351.

Guccione, M.J., Van Arsdale, R.B., Hehr, L.H., Ewen, C.R., and Meyers, R. (1993). *Origin and age of the "Sunk Lands" Big Lake, Arkansas*. Report submitted to the U.S. Geological Survey, Award 14-08-0001-61997: Washington, D.C., p. 44.

Guccione, M.J., Miller, J.Q., and Van Arsdale, R.B. (1994). Amount and timing of deformation near the St. Francis "Sunklands," northeastern Arkansas. *Geol. Soc. Am. Abstr. Prog.*, **26**(1), 7.

Guiseppe, A.C. and Heller, P.L. (1998). Long-term river response to regional doming in the Price River formation, central Utah. *Geology*, **26**, 239–42.

Gupta, S. (1997). Himalayan drainage patterns and the origin of fluvial megafans in the Ganges foreland basin. *Geology*, **25**, 11–14.

Gutenberg, B. (1941). Changes in sea level, postglacial uplift, and mobility of the earth's interior. *Geol. Soc. Am. Bull*, **52**, 721–72.

Harbor, D.J. (1998). Dynamic equilibrium between an active uplift and the Sevier River, UT. *Geol. Soc. Am. Bull.*, **106**, 181–94.

Harbor, D.J., Schumm, S.A., and Harvey, M.D. (1994). Tectonic control of the Indus River in Sindh, Pakistan. In Schumm, S.A. and Winkley, B.R. (eds.) *The variability of large alluvial rivers*. American Society of Civil Engineers Press, New York, pp. 161–75.

Harrison, T.S. (1927). Colorado–Utah Salt Domes. *J. Geol.*, **11**, 111–35

Harvey, M.D. and Schumm, S.A. (1994). Alabama River: variability of overbank flooding and deposition. In Schumm, S.A. and Winkley, B.R. (eds.) *The variability of large alluvial rivers*. American Society Civil Engineers Press, New York, pp. 313–37.

Harwood, D.S. and Helley, E.J. (1987). *Late Cenozoic tectonism of the Sacramento Valley, California*. U.S. Geol. Survey Prof. Paper 1359, 46 pp.

Hawley, J.W. and Wilson, W.E. (1965). *Quaternary geology of the Winnemucca area, Nevada*. Univ. Nevada Desert Research Institute Technical Report 5, 66 pp.

Hazel, J.E. Jr (1991). Alluvial architecture of the Upper Triassic Chinle Formation, Cane Creek anticline, Canyonlands, UT. M.S. Thesis, Northern Arizona Univ., Flagstaff, 149 pp.

Hazel, J.E. Jr (1994). Sedimentary response to intrabasinal salt tectonism in the Upper Triassic Chinle Formation, Paradox Basin, UT. *U.S. Geol. Surv. Bull.*, **2000–F**, F1–34.

Heller, P.L., Angevine, C.L., Winslow, N.S., and Paola, C. (1988). Two-phase stratigraphic model of foreland-basin sequences. *Geology*, **16**, 501–4.

Heller, P.L. and Paola, C. (1996). Downstream changes in alluvial architecture: an exploration of controls on channel-stacking patterns. *J. Sed. Res.*, **66**(2), 297–306.

Heyl, A.V. and McKeown, F.A. (1978). *Preliminary seismotectonic map of central Mississippi Valley and environs*. U.S. Geol. Survey Misc. Field Studies Map, MF-1011.

Hildenbrand, T.G., Kane, M.F. and Hendricks, J.D. (1982). Magnetic basement in the upper Mississippi Embayment region – a preliminary report. In *Investigations of the New Madrid, MO, Earthquake Region*: U.S. Geol. Survey Prof. Paper 1236, pp. 39–53.

Hobday, D.K., Woodruff, C.M. Jr, and McBride, M.W. (1981). *Paleotopographic and structural controls on non-marine sedimentation of the Lower Cretaceous Antlers Formation and correlatives, north Texas and southeastern Oklahoma*. Society of Sedimentary Geology. Special Publ. 31, pp. 71–87.

Holbrook, J.M. (1992). Developmental and sequence-stratigraphic analysis of Lower Cretaceous sedimentary systems in the southern part of the United States Western Interior: interrelationships between eustasy, local tectonics, and depositional environments: Ph.D. Dissertation, Indiana Univ., Bloomington, IN, 271 pp.

Holbrook, J.M. (1996a). Complex fluvial response to low gradients at maximum regression: a genetic link between smooth sequence-boundary morphology and architecture of overlying sheet sandstone. *J. Sed. Res.*, **66**(4), 713–22.

Holbrook, J.M. (1996b). Structural noise in seemingly undeformed intraplate regions: implications from welts raised in a shattered Aptian/Albian U.S. Western Interior. *Theophrastus' Contributions*, **1**, 87–94.

Holbrook, J.M. and White, D.W. (1999). Controls on lithofacies distribution and sequence architecture by low-relief intraplate uplift: examples from the Lower Cretaceous of northeastern New Mexico. In Shanley, K.W. and McCabe, P.J. (eds.) *Relative role of eustasy, climate, and tectonism in continental rocks*. Society Sedimentary Geology. Special Publ. 59 (in press).

Holbrook, J.M. and Wright Dunbar, R. (1992). Depositional history of Lower Cretaceous strata in northeastern New Mexico: implications for regional tectonics and depositional sequences. *Geol. Soc. Am. Bull.*, **104**(7), 802–13.

Holdahl, S.R. and Morrison, N.L. (1974). Regional investigation of vertical crustal movements in the U.S., using precise relevelings and mareograph data. *Tectonophysics*, **23**, 373–90.

Holmes, A. (1965). *Principles of physical geology*. London, T. Nelson, 1288 pp.

Holmes, D.A. (1968). The recent history of the Indus. *Geog. J.* **134**(3), 367–82.

Holz, K.R., Baker, V.R., Sutton, S.M., and Penteado-Orellana, M.M. (1979). *South American river morphology and hydrology*. NASA Special Pub. 412, Apollo-Soyuz Test Project Summary Science Report, **II**, 545–94.

Hopkins, J.C. (1985). Channel-fill deposits formed by aggradation in deeply scoured, superimposed distributaries of the Lower Kootenai Formation (Cretaceous). *J. Sed. Petrol.*, **55**(1), 42–52.

Horváth, F. (1988). Neotectonic behavior of the Alpine-Mediterranean region. *Am. Assoc. Petrol. Geol. Mem.*, **45**, 49–55.

Horváth, F. (1993). Towards a mechanical model for the formation of the Pannonian Basin. *Tectonophysics*, **226**, 333–57.

Howard, A. D. (1967). Drainage analysis in geologic interpretation: a summary. *Am. Assoc. Petrol. Geol.*, **51**, 2246–59.

Huang, W. (1993). Morphologic patterns of stream channels on the active Yishi Fault, southern Shandong Province, Eastern China: implication for repeated great earthquakes in the Holocene. *Tectonophysics*, **219**, 283–304.

Hunt, C. B. (1979). The Great Basin, an overview and hypotheses of its history. *RMAG-UGA Proceedings of the 1979 Basin and Range Symposium*, pp. 1–9.

Hunt, C. B. and Mabey, D. R. (1966). *Stratigraphy and structure, Death Valley, California*. U.S. Geol. Survey Prof. Paper 494–A, 156 pp.

Hunt, C. B., Robinson, T. W., Bowles, W. A., and Washburn, A. L. (1966). *Hydrologic basin, Death Valley, California*. U.S. Geol. Surv. Prof. Paper 494–B, 133 pp.

Huntoon, P. W. (1982). The meander anticline, Canyonlands, UT, an unloading structure resulting from horizontal gliding on salt. *Geol. Soc. Am. Bull.*, **93**, 941–50.

Huntoon, P. W. (1988). Late Cenozoic gravity tectonic deformation related to the Paradox salts in the Canyonlands area of Utah. *Utah Geol. and Mineral Survey Rep.* 122, pp. 79–93.

Huntoon, P. W. and Elston, D. P. (1979). *Origin of the river anticlines, central Grand Canyon, AZ*. U.S. Geol. Survey Prof. Paper 1126–A, 9 pp.

Irion, G. (1984). Sedimentation and sediments of Amazonian rivers and evolution of the Amazonian landscape since Pliocene times. In: Sioli, H. (ed.) *The Amazon, Limnology and Landscape Ecology of a Mighty Tropical River and its Basin*, pp. 201–13, W. Junk, Dordrecht.

Jackson, J., Norris, R., and Youngson, J. (1996). The structural evolution of active fault and fold systems in central Otago, New Zealand: evidence revealed by drainage patterns: *J. Struct. Geol.*, **18**, 217–34.

Jackson, M. P. A. (1982). *Fault tectonics of the East Texas Basin*. Univ. Texas, Bureau of Economic Geology Geol. Circular 82–4, 31 pp.

Jackson, M. P. A. and Talbot, C. J. (1986). External shapes, strain rates, and dynamics of salt structures. *Geol. Soc. Am. Bull.*, **97**, 305–23.

Jefferson, M. (1907). Lateral erosion on some Michigan rivers. *Geol. Soc. Am. Bull.*, **18**, 333–50.

Jervey, M. T. (1988). Quantitative geologic modeling of siliciclastic rock sequences and their seismic expression. In Wilgus, C. K., Hastings, B. S., Kendall, C. G. st. C., Posamentier, H. W., Ross, C. A., and Van Wagoner, J. C. (eds.) *Sea level change – an integrated approach*. Society Economic Paleontologists and Mineralogists, Special Pub. 42, pp. 47–69.

Jibson, R. W. and Keefer, D. K. (1989). Statistical analysis of factors affecting landslide distribution in the New Madrid seismic zone, Tennessee and Kentucky. *Eng. Geol.*, **27**, 509–42.

Jibson, R. W. and Keefer, D. K. (1993). Analysis of the seismic origin of landslides – examples from the New Madrid seismic zone. *Geol. Soc. Am. Bull.*, **105**, 521–36.

Jin, D. (1983). Unpublished report on experimental studies. Colorado State Univ., 15 pp.

Jin, D. and Schumm, S.A. (1987). A new technique for modeling river morphology. In Gardner, V. (ed.). *International Geomorphology*. Wiley, Chichester, pp. 681–90.

Johnson, E.A. (1989). Depositional environments and tectonic controls on the coal-bearing Lower to Middle Jurassic Yan'an Formation, southern Ordos Basin, China. *Geology*, **17**, 1123–6.

Johnson, S.Y. and Alam, A.M.N. (1991). Sedimentation and tectonics of the Sylheth trough, Bangladesh. *Geol. Soc. Am. Bull.*, **103**, 1513–27.

Johnston, A.C. (1982). A major earthquake zone on the Mississippi. *Sci. Am.*, **246**, 60–8.

Johnston, A.C. and Shedlock, K.M. (1992). Overview of research in the New Madrid seismic zone. *Seismol. Res. Lett.*, **63**(3), 193–208.

Joó, I. (1992). Recent vertical surface movements in the Carpathian Basin. *Tectonophysics*, **202**, 129–34.

Jordan, T.E. (1981). Thrust loads and foreland basin evolution, Cretaceous western United States. *Am. Assoc. Petrol. Geol. Bull.*, **15**, 2500–20.

Jorgensen, D.W. (1990). Adjustment of alluvial river morphology and process to localized active tectonics. Unpublished Ph.D. dissertation, Colorado State Univ., 240 pp.

Jorgensen, D.W., Harvey, M.D., Schumm, S.A., and Flam, L. (1993). Morphology and dynamics of the Indus River: implications for the Mohen jo Daro site. In Shroder, J.F., Jr (ed.) *Himalaya to the sea*. Routledge, London, pp. 288–326.

Jurkowski, G., Ni, J., and Brown, W. (1984). Modern parching of the Gulf Coastal Plains. *J. Geophys. Res.*, **89**, 6247–55.

Kafri, V. (1969). Recent crustal movements in northern Israel. *J. Geophys. Res.*, **74**, 4246–58.

Kaizuka, S. (1967). Rate of folding in the Quaternary and the present. *Geog. Rep. Tokyo Metropol. Univ.*, **2**, 1–10.

Kazmi, A.H. (1979). Active faults in Pakistan, Geodynamics of Pakistan. In A. Farah and K.A. DeJong (eds.) *Geological Survey of Pakistan*, Quetta.

Keller, E.A. and Pinter, N. (1996). *Active tectonics*. Prentice-Hall, Upper Saddle River, 338 pp.

Keller, E.A., Bonkowski, M.S., Korsch, R.J., and Schlemon, R.J. (1982). Tectonic geomorphology of the San Andreas fault zone in the southern Indio Hills, Coachella Valley, CA. *Geol. Soc. Am. Bull.*, **93**, 46–56.

Keller, E.A., Zepeda, R.L., Rockwell, T.K., Ku, T.L., and Dinklage, W.L. (1998). Active tectonics at Wheeler Ridge, Southern San Joaquin Valley, CA. *Geol. Soc. Am. Bull.*, **110**, 298–310.

Kent, B.H. (1986). Evolution of thick coal deposits in the Powder River Basin, northeastern Wyoming. *Geol. Soc. Am. Special Paper* **210**, 105–22.

Kerr, R.A. (1995). Earth's surface may move itself. *Science*, **269**, 1214–15.

King, G. and Stein, R. (1983). *Surface folding, river terrace deformation rate and earthquake repeat time in a reverse faulting environment: in the 1983 Coalinga, California Earthquake*. California Division of Mines Publ. 66, pp. 1–12.

King, G., Bailey, G., and Sturdy, D. (1994). Active tectonics and human survival strategies. *J. Geophys. Res.*, **99**, 20 063–8.

King, P.B. and Schumm, S.A. (1980). *The physical geography (geomorphology) of William Morris Davis*. Geo. Books, Norwich, 217 pp.

Kirk, A.R. and Condon, S.M. (1986). Structural control of sedimentation patterns and the distribution of uranium deposits in the Westwater Canyon member of the Morrison formation, northwestern New Mexico – A subsurface study. In Turner-

Peterson, C.E., Santos, E.S., and Fishman, N.S., (eds.) *A basin analysis case study: the Morrison formation Grants Uranium Region New Mexico.* American Association of Petroleum Geology, Studies in Geology 22, pp. 105–43.

Knighton, A.D. and Nanson, G.C. (1993). Anastomosis and the continuum of channel pattern. *Earth Surface Processes and Landforms*, **18**, 613–25.

Kowalski, W.C. and Radzikoska, H. (1968). The influence of neotectonic movements on the formation of alluvial deposits and its engineering-geological estimation. *Proc. 23rd International Geology Congress, Prague, Czech.*, Section 12, pp. 197–203.

Kraus, M.J. (1996). Avulsion deposits in Lower Eocene alluvial rocks, Bighorn Basin, Wyoming. *J. Sed. Res.*, **66**(2), 354–63.

Kraus, M.J. and Middleton, L.T. (1987). Contrasting architecture of two alluvial suites in different structural settings. In Ethridge, F.G., Flores, R.M., and Harvey, M.D., (eds.) *Recent developments in fluvial sedimentology.* Society of Economic Paleontologists and Mineralogists, Special Pub. 39, pp. 253–62.

Krzyszkowski, D. and Stachura, R. (1997). Neotectonically controlled fluvial features, Walbrzych Upland, Middle Sudeten Mts., southwestern Poland. *Geomorphology*, **22**, 73–91.

Kuenzi, W.D. (1966). Tertiary stratigraphy in the Jefferson River basin, Montana. Ph.D. dissertation, Bozeman, University of Montana, 390 pp.

Kvale, E.P. and Vondra, C.F. (1993). *Effects of relative sea-level changes and local tectonics on a Lower Cretaceous fluvial to transitional marine sequence, Bighorn Basin, Wyoming, USA.* International Association of Sedimentologists Special Publ. 17, pp. 383–99.

Lamotte, S. (1990). Fluvial dynamics and succession in the lower Ucayali River Basin, Peruvian Amazonia. *For. Ecol. Man.*, **33/34**, 141–56.

Lane, E.W. (1955). Design of stable channels. *Am. Soc. Civ. Eng. Trans.*, **120**, 1–34.

Lane, E.W. (1957). *A study of the shape of channels formed by natural streams flowing in erodible material.* U.S. Army Corps of Engineers, Missouri River Division, Omaha, Nebraska, Sediment Series 9.

Lattman, L.H. (1959). Geomorphology: new tool for finding oil. *Oil Gas J.*, **57**, 231–6.

Laurent, H. (1985). El pre-Cretaceo en el oriente peruana: su distribucon y sus ragos estructurales. *Boletin de la Sociedad Geologia del Peru*, **74**, 33–59.

Laurent, H. and Pardo, A. (1975). Ensayo de interpretacion del basamento del Noroiente Peruano. *Boletin de la Sociedad Geologica del Peru*, **45**, 25–48.

Lawler, D.M. (1986). River bank erosion and the influence of frost: a statistical examination. *Inst. Br. Geog. Trans., New Series*, **11**, 227–42.

Lawrence, D.A. and Williams, B.P.J. (1987). Evolution of drainage systems in response to Acadian deformation: the Devonian Battery Point Formation, Eastern Canada. In Ethridge, F.G., Flores, R.M., and Harvey, M.D. (eds.), *Recent developments in fluvial sedimentology.* Society of Economic Paleontologists and Mineralogists, Special Publ. 39, pp. 287–300.

Lawson, A.C. (1912). The recent fault scarp at Genoa, NV. *Seismol. Soc. Am. Bull.*, **2**, 193–200.

Leary, P.C., Malin, P.E., Strelitz, R.A., and Henrey, T.L. (1981). Possible tilt phenomena observed as water level anomalies along the Los Angeles aqueduct. *Geophys. Res. Lett.*, **8**, 225–8.

Lecarpentier, C. and Motti, E. (1969). Reconnaissance hydro-géomorphologique du Pachitea (Amazonie peruvienne). Ministère de l'Education Nationale, Comité des Travaux Historiques et Scientifiques. *Bull. de la Section de Geographie*, **LXXX**, 485–513.

Leeder, M. R. (1993). Tectonic controls upon drainage development, river channel migration and alluvial architecture: implications for hydrocarbon reservoir development and characterization. In North, C. P. and Prosser, D. J. (eds.) *Characterization of fluvial and aeolian reservoirs*. Geological Society of America Special Publ. 73, pp. 7–22.

Leeder, M. R. and Alexander, J. (1987). The origin and tectonic significance of asymmetrical meanderbelts. *Sedimentology*, **34**, 217–26.

Leeder, M. R. and Gawthorpe, R. L. (1987). Sedimentary models for extensional tilt block/half-graben basins. In Coward, M. P., Dewey, J. F., and Hancock, P. L. (eds.) *Continental extensional tectonics*. Geological Society of London Special Publ. 28, pp. 139–52.

Leeder, M. R. and Jackson, J. A. (1993). The interaction between normal faulting and drainage in active extensional basins with examples from the western United States and central Greece. *Basin Res.*, **5**, 79–102.

Leeder, M. R., Mack, G. H., Peakall, J. and Salyards, S. L. (1996). First quantitative test of alluvial stratigraphic models: southern Rio Grande rift, New Mexico. *Geology*, **24**, 87–90.

Lees, G. M. (1955). Recent earth movements in the Middle East. *Geol. Rundschau*, **43**, 221–6.

Leopold, L. B. and Bull, W. B. (1979). Base level, aggradation, and grade. *Am. Philos. Soc. Proc.*, **123**(3), 168–202.

Leopold, L. B. and Wolman, M. G. (1957). *River channel patterns; braided, meandering, and straight*. U. S. Geol. Surv. Prof. Paper 282–B.

le Roux, J. P. (1994). Persistence of topographic features as a result of non-tectonic processes. *Sed. Geol.*, **89**, 33–42.

Lloyd, M. J. (1994). Sediment provenance studies in the Pyrenean foreland basin, Aragon, Spain. Ph. D. Dissertation, Cambridge University, 325 pp.

Lorenz, J. C. (1982). Lithospheric flexture and the history of the Sweetgrass Arch northwestern Montana. In Powers, R. B. (ed.), *Geologic studies of the Cordilleran thrust belt*. Rocky Mountain Association of Geologists, pp. 77–89.

Lundin, E. R. (1989). Thrusting of the Claron Formation, the Bryce Canyon region, UT. *Geol. Soc. Am. Bull.*, **101**, 1038–50.

Luzietti, E. A., Kanter, L. R., Schweig, E. S., III, Shedlock, K. M., and Van Arsdale, R. B. (1992). Shallow deformation along the Crittenden County fault zone near the southeastern boundary of the Reelfoot Rift, northeast Arkansas. *Seismol. Res. Lett.*, **63**, 263–76.

McCalpin, J. P. (1996). *Paleoseismology*. Academic Press, San Diego, 583 pp.

McCarthy, T. S., Ellery, W. N., and Stanistreet, I. G. (1992). Avulsion mechanisms on the Okavango fan, Botsawana: the control of a fluvial system by vegetation. *Sedimentology*, **39**, 779–95.

McGinnis, L. D. (1963). *Earthquakes and crustal movement as related to water load in the Mississippi valley region*. Illinois State Geol. Survey Circular 344, 20 pp.

McKeown, F. A., Jones-Cecil, M., Askew, B. L., and McGrath, M. B. (1988). *Analysis of stream-profile data and inferred tectonic activity, eastern Ozark Mountains region*. U. S. Geol. Survey Bull., 1807, 39 pp.

McKeown, F. A., Hamilton, R. M., Diehl, S. F., and Glick, E. E. (1990). Diapiric origin of the Blytheville and Pascola Arches in the Reelfoot Rift, east-central United States: relation to New Madrid seismicity. *Geology*, **18**, 1158–62.

Machette, M.N. (1978). Dating Quaternary faults in the southwestern United States by using buried calcareous soils. *J. Res. U.S. Geol. Surv.*, **6**, 369–81.

Machida, T. (1960). Geomorphological analysis of terrace plains-fluvial terraces along the River Kuji and River Ara, Kanto District, Japan. Tokyo Kyoiku Daigaku (Tokyo Univ. of Education). *Sci. Rep. Sect. C*, **7**, 137–94.

Mack, G.H. and James, W.C. (1993). Control of basin symmetry on fluvial lithofacies, Camp Rice and Palomas Formations (Plio-Pleistocene), southern Rio Grande rift, USA. In Marzo, M. and Puigdefábregas, C. (eds.) *Alluvial Architecture*. International Association of Sedimentologists Special Publ. 17, pp. 439–49.

Mack, G.H. and Seager, W.R. (1990). Tectonic control on facies distribution of the Camp Rice and Palomas Formations (Pliocene-Pleistocene) in the southern Rio Grande rift. *Geol. Soc. Am. Bull.*, **102**, 45–53.

Mackey, S. and Bridge, J.S. (1995). Three-dimensional model of alluvial stratigraphy: theory and application. *J. Sed. Res.*, **B65**(1), 7–31.

Mackin, J.H. (1948). Concept of the graded river. *Geol. Soc. Am. Bull.*, **59**, 463–512.

Maizels, J. (1979). Proglacial aggradation and changes in braided channel patterns during a period of glacier advance: an alpine example. *Geografiska Annaler, A.*, **61**, 87–101.

Markewich, H.W. (1985). Geomorphic evidence for Pliocene–Pleistocene uplift in the area of the Cape Fear Arch, North Carolina. In Morisawa, M. and Hack, J.T. (eds.) *Tectonic geomorphology*. Allen and Unwin, Boston, MA, pp. 279–97.

Marple, R.T. (1994). Discovery of a possible seismogenic fault system beneath the coastal plain of South and North Carolina from an integration of river morphology and geological and geophysical data. Unpublished Ph.D. dissertation, University of South Carolina, 354 pp.

Marple, R.T. and Talwani, P. (1993). Evidence of possible tectonic upwarping along the South Carolina coastal plain from an examination of river morphology and elevation data. *Geology*, **21**, 651–4.

Marshak, S. and Paulson, T. (1996). Midcontinent U.S. fault and fold zones: a legacy of Proterozoic intracratonic extensional tectonism? *Geology*, **24**(2), 151–4.

Martin, R.G. (1978). Northern and eastern Gulf of Mexico continental margin: stratigraphic and structural framework. In Bouma, A.H., Moore, G.T., and Coleman, J.M. (eds.) *Framework, facies, and oil-trapping characteristics of the upper continental margin*. American Association of Petroleum Geology, Studies in Geology No. 7, pp. 21–42.

Martins-Neto, M.A. (1994). Braided sedimentation in a Proterozoic rift basin: the São João da Chapada Formation, southeastern Brazil. *Sed. Geol.*, **89**, 219–39.

Mather, A.E. (1993). Basin inversion: some consequences for drainage evolution and alluvial architecture. *Sedimentology*, **40**, 1069–89.

Megard, F (1984). The Andean orogenic period and its major structures in central and northern Peru. *Geol. Soc. Lond. J.*, **141**, 893–900.

Melton, F.A. (1959). Aerial photographs and structural geology. *J. Geol.*, **67**, 351–70.

Melvin, J. (1993). Evolving fluvial style in the Kekiktuk Formation (Mississippian), Endicott Field area, Alaska: base level response to contemporaneous tectonism. *Am. Assoc. Petrol. Geol. Bull.*, **77**(10), 1723–44.

Merritts, D. and Hesterberg, T. (1994). Stream networks and long-term surface uplift in the New Madrid Seismic Zone. *Science*, **265**, 1081–4.

Merritts, D. and Vincent, K.R. (1989). Geomorphic response of coastal streams to low, intermediate, and high rates of uplift, Mendocino triple junction region, northern California. *Geol. Soc. Am. Bull.*, **101**, 1373–88.

Merritts, D.J., Vincent, K.R., and Wohl, E.E. (1994). Long river profiles, tectonism, and eustasy: a guide to interpreting fluvial terraces. *J. Geophy. Res.*, **99**, 14031–50.

Meyers, J.H., Suttner, L.J., Furer, L.C., May, M.T., and Soreghan, M.J. (1992). Intrabasinal tectonic control on fluvial sandstone bodies in the Cloverly Formation (Early Cretaceous), west-central Wyoming, USA. *Basin Res.*, **4**, 315–33.

Miall, A.D. (1976). Paleocurrent and paleohydrologic analysis of some vertical profiles through a Cretaceous braided stream deposit, Banks Island, Arctic Canada. *Sedimentology*, **23**, 459–84.

Miall, A.D. (1978). Tectonic setting and syndepositional deformation of molasse and other nonmarine-paralic sedimentary basins. *Can. J. Earth Sci.*, **15**, 1613–32.

Miall, A.D. (1985). Architectural-element analysis: a new method of facies analysis applied to fluvial deposits. *Earth Sci. Rev.*, **22**, 261–308.

Miall, A.D. (1991). Hierarchies of architectural units in clastic rocks, and their relationship to sedimentation rate. In Miall, A.D. and Tyler, N. (eds.) *The three-dimensional facies architecture of terrigenous clastic sediments, and its implications for hydrocarbon discovery and recovery*. Society of Economic Paleontologists and Mineralogists, Concepts in Sedimentology and Paleontology 3, pp. 6–12.

Miall, A.D. (1996). *The geology of fluvial deposits – sedimentary facies, basin analysis, and petroleum geology*. Springer-Verlag, Berlin, 582 pp.

Mike, K. (1975). Utilization of the analysis of ancient river beds for the detection of Holocene crustal movements. *Tectonophysics*, **29**, 359–68.

Mississippi River Flood Control Association (1927). Unpublished report on 1927 flood.

Morgan, J.P. and McIntire, W.G. (1959). Quaternary geology of the Bengal Basin, East Pakistan and India. *Geol. Soc. Am. Bull.*, **70**, 319–42.

Morisawa, M. (1975). Tectonics and geomorphic models. In Melhorn, W.N. and Flemal, R.C. (eds.) *Theories of landform development*. Publications in Geomorphology, SUNY Binghamton, NY, pp. 199–216.

Morisawa, M. and Hack, J.T. (eds.) (1985). *Tectonic geomorphology*. Allen & Unwin, Boston, MA.

Moseley, M.E. (1983). The good old days were better: agrarian collapse and tectonics. *Am. Anthropol.*, **85**, 773–99.

Muehlberger, W.R. (1979). *The Embudo fault between Pilar and Arroyo Hondo, New Mexico*. New Mexico Geol. Soc. Guidebook, 30th Field Conf., pp. 77–82.

Müller, B., Zoback, M.L., Fuch, K., Mastin, L., Gregersen, S., Pavoni, N., Stephansson, O., and Ljunggren, C. (1992). Regional patterns of tectonic stress in Europe. *J. Geophys. Res.*, **97** (B8), 11783–803.

Murray, G.E. (1961). *Geology of the Atlantic and Gulf Coast province of North America*. Harper, NY, 692 pp.

Myers, W.B. and Hamilton, W. (1964). *Deformation accompanying the Hebgen Lake earthquake of August 17, 1959*. U.S. Geol. Survey Prof. Paper 435–I, pp. 55–98.

Nagymarosy, A. and Báldi-Beke, M. (1993). The Szolnok unit and its probable paleogeographic position. *Tectonophysics*, **226**, 457–70.

Nakayama, K. (1994). Stratigraphy and paleogeography of the Upper Cenozoic Tokai

Group around the east coast of Ise Bay, Central Japan. *J. Geosci., Osaka City Univ.*, **37**, 77–143.

Nakayama, K. (1996). Depositional models for fluvial sediments in an intra-arc basin: an example from the Upper Cenozoic Todai Group in Japan. *Sed. Geol.*, **101**, 193–211.

Nanson, G. C. (1980a). A regional trend to meander migration. *J. Geol.*, **88**, 100–8.

Nanson, G. C. (1980b). Point bar and floodplain formation of the meandering Beaton River, northeastern British Columbia, Canada. *Sedimentology*, **27**, 3–29.

Nash, D. B. (1984). Morphologic dating of fluvial terrace scarps and fault scarps near west Yellowstone, MT. *Geol. Soc. Am. Bull.*, **95**, 1413–24.

Neef, E. (1966). Geomorphologische Möglichkeiten für feststellung junger Erdkrustenbewegungen. *Geologie (Berlin)*, **15**, 97–101.

Neev, D. and Emery, K. O. (1995). *The destruction of Sodom, Gomorrah, and Jerico*. Oxford University Press, Oxford, 175 pp.

Nichols, G. J. (1987). Structural controls of fluvial distributary systems – the Luna System, northern Spain. In Ethridge, F. G., Flores, R. M., and Harvey, M. D. (eds.) *Recent developments in fluvial sedimentology*. Soc. Economic Paleontologists and Mineralogists, Special Pub. 39, pp. 269–77.

Nunn, J. A. (1985). State of stress in the northern Gulf Coast. *Geology*, **13**, 429–32.

Nuttli, O. W. (1982). Damaging earthquakes of the central Mississippi valley. In *Investigations of the New Madrid, MO, earthquake region*. U.S. Geol. Survey Prof. Paper 1236, pp. 15–20.

Obermeier, S. F. (1989). *The New Madrid earthquakes: an engineering-geologic interpretation of relict liquefaction features*. U.S. Geol. Survey, Prof. Paper 1336–B, 114 pp.

O'Connell, D. R., Bufe, C. G., and Zoback, M. D. (1982). *Microearthquakes and faulting in the area of New Madrid, MO, Reelfoot Lake, TN*. U.S. Geol. Survey Prof. Paper 1236, pp. 31–8.

Odum, J. K., Stephenson, W. J., Shedlock, K. M., and Pratt, T. L. (1998). Near-surface structural model for deformation associated with the February 7, 1812, New Madrid, Missouri earthquake. *Geol. Soc. Am. Bull.*, **110**, 149–62.

Oldham, R. D. (1926). The Cutch earthquake of 16 June 1819 with a revision of the great earthquake of 12 June 1897. *Geol. Surv. of Ind. Mem.* **46**, 1–77.

Ollier, C. D. (1981). *Tectonics and landforms*. Longman Group Limited, Harlow, England, 324 pp.

Orme, C. D. (1980). Marine terraces and Quaternary tectonism, northwest Baja California, Mexico. *Phys. Geog.*, **1**, 138–61.

Osborn, G. and du Toit, C. (1991). Lateral plantation of rivers as a geomorphic agent. *Geomorphology*, **4**, 249–60.

Otvos, E. G. (1981). Tectonic lineaments of Pliocene and Quaternary shorelines, northeast Gulf Coast. *Geology*, **9**, 398–404.

Ouchi, S. (1983). Response of alluvial rivers to slow active tectonic movement. Ph.D. dissertation, Colorado State Univ., Fort Collins, CO, 205 pp.

Ouchi, S. (1985). Response of alluvial rivers to slow active tectonic movement. *Geol. Soc. Am. Bull.*, **96**, 504–15.

Paola, C. (1988). Subsidence and gravel transport in alluvial basins. In Kleinspehn, K. and Paola, C. (eds.) *New perspectives in basin analysis*. Springer-Verlag, NY, pp. 231–43.

Pardee, J. T. (1950). Late Cenozoic block faulting in western Montana. *Geol. Soc. Am. Bull.*, **61**, 359–406.

Pardo, A. (1982). *Characteristicas estructurales de la faja subandina del norte del Peru*. Symposium Exploracion Petrolera and las Cuencas Subandinas de Venzuela, Colombia, Ecuador y Peru, Bogota.

Pashin, J.C. (1994). Coal-bed geometry and synsedimentary detachment folding in Oak Grove coalbed methane field, Black Warrior Basin, Alabama. *Am. Assoc. Petrol. Geol. Bull.*, **78**(6), 960–80.

Peakall, J. (1996). The influence of lateral ground-tilting on channel morphology and alluvial architecture. Ph.D. Dissertation, Univ. of Leeds, Leeds, 320 pp.

Peakall, J. (1998). Axial river evolution in response to half-graben faulting: Carson River, Nevada. *Geol. Soc. Am. Bull.* **68**, 788–99.

Penck, W. (1953). *Morphological analysis of landforms*. MacMillan, London, 429 pp.

Penick, J., Jr. (1981). *The New Madrid earthquakes of 1811–1812*: Univ. Missouri Press, Columbia, MO, 181 pp.

Peterson, F. (1984). Fluvial sedimentation on a quivering craton: influence of slight crustal movements on fluvial processes, Upper Jurassic Morrison Formation, western Colorado Plateau. *Sed. Geol.*, **38**, 21–49.

Phillips, L. and Schumm S.A. (1987). Effect of regional slope on drainage networks. *Geology*, **15**, 813–16.

Pivnik, D.A. and Johnson, G.D. (1995). Depositional response to Pliocene–Pleistocene foreland partitioning in northwest Pakistan. *Geol. Soc. Am. Bull.*, **107**(8), 895–922.

Plafker, G. (1964). Oriented lake and lineaments of northeastern Bolovia. *Geol. Soc. Am. Bull.*, **75**, 503–22.

Plafker, G. (1969). *Tectonics of the March 27, 1964 Alaska earthquake*. U.S. Geol. Survey Prof. Paper 543–I, 74 pp.

Playfair, J. (1802). *Illustrations of the Huttonian theory of the Earth*. Cadell and Davis, London 528 pp.

Poland, J.F. (1972). Subsidence and its control. *Am. Assoc. Petrol Geol., Mem.,* **18**, 50–71.

Poland, J.F. (1976). Land subsidence stopped by artesian head recovery, Santa Clara Valley, CA. *Int. Assoc. Hydrol. Sci.*, **121**, 124–32.

Poland, J.F. (1981). Subsidence in United States due to ground-water withdrawal. *J. Irrig. Drain. Div., Am. Soc. Civil Eng.*, **107**, 115–35.

Poland, J.F. and Green, J.H. (1962). *Subsidence in the Santa Clara Valley, CA – a progress report*. U.S. Geol. Survey, Water-Supply Paper 1619–C, 16 pp.

Poland, J.F., Lofgren, B.E., Ireland, R.L., and Pugh, R.G. (1975). *Land subsidence in the San Joaquin Valley as of 1972*. U.S. Geol. Survey Prof. Paper 437–H, 78 pp.

Posamentier, H.W. and Vail, P.R. (1988). Eustatic controls on clastic deposition II – sequence and systems tract models. In Wilgus, C.K., Hastings, B.S., Kendall, C.G.st.C., Posamentier, H.W., Ross, C.A., and Van Wagoner, J.C. (eds.) *Sea level change – an integrated approach*. Society of Economic Paleontologists and Mineralogists, Special Pub. 42, p. 125–54.

Potter, D.B., Jr. and Gill, G.E. (1978). Valley anticlines of the Needles District, Canyonlands National Park, UT. *Geol. Soc. Am. Bull.*, **89**, 952–60.

Potter, P.E. (1978). Significance and origin of big rivers. *J. Geol.*, **86**, 13–33.

Price, R.C. and Whetstone, K.N. (1977). Lateral stream migration as evidence for geologic structures in the eastern Gulf Coastal Plain. *Southeast. Geol.*, **18**, 129–48.

Prokopovich, N.P. (1983). Neotectonic movement and subsidence caused by piezometric decline. *Bull. Assoc. Eng. Geol.*, **20**, 393–404.

Radulescu, G. (1962), Deciphering of tectonic movement in the Quaternary territory of Rumania by means of geomorphological methods. *Prob. de Geog.*, 9–19.

Räsänen, M.E., Salo, J.S., and Kaliola, R.J. (1987). Fluvial perturbance in the western Amazon river basin: regulation by long term sub-Andean tectonics. *Science*, **238**, 1398–401.

Räsänen, M.E., Neller, R., Salo, J., and Jungmert, H. (1991). Recent and ancient fluvial depositional systems in the Andean forelands of the Peruvian Amazon. In *History of the fluvial and alluvial landscapes of the western Amazon Andean Forelands*. Annales Universitatis Turkuensis, Ser. A, **75** (3) 1–22.

Read, W.A. and Dean, J.M. (1982). Quantitative relationships between numbers of fluvial cycles, bulk lithological composition and net subsidence in a Scottish Namurian basin. *Sedimentology*, **29**, 181–200.

Reid, J.B., Jr. (1992). The Owens River as a tiltmeter for Long Valley Caldera, CA. *J. Geol.*, **100**, 353–63.

Reilinger, R.E. and Oliver, J.E. (1976). Modern uplift associated with a proposed magma body in the vicinity of Socorro, NM. *Geology*, **4**, 583–6.

Reilinger, R.E., Oliver, J.E., Sanford, A. and Balazs, E. (1980). New measurements of crustal doming over the Socorro magma body, New Mexico. *Geology*, **8**, 291–5.

Reshkin, M. (1963). Geomorphic history of the Jefferson Basin, Jefferson, Madison, and Silverbow Counties, Montana. Ph.D. dissertation, Indiana Univ., Bloomington, 146 pp.

Reynolds, M.W. (1979). *Character and extent of Basin Range faulting, western Montana and east-central Idaho*. 1979 Basin and Range Symposium, pp. 185–93.

Rhea, S. and Wheeler, R.L. (1994). Folio of seismotectonic maps of the New Madrid area, southeastern Missouri and adjacent states. U.S. Geol. Survey Misc. Field Studies Map MF-2664, scale 1:250 000.

Riad, S. (1977). Shear zones in north Egypt interpreted from gravity data. *Geophysics*, **42**, 1207–14.

Riba, O. (1976). Syntectonic unconformities of the Alto Cardener, Spanish Pyrenees: a genetic interpretation. *Sed. Geol.*, **15**, 213–33.

Richards, K., Chandra, S., and Friend, P. (1993). Avulsive channel systems: characteristics and examples. In Best, J.L. and Bristow, C.S., (eds.) *Braided Rivers*. Geological Society of London, Special Publ. 75, pp. 195–203.

Rockwell, T.K. (1988). Neotectonics of the San Cayetano fault, Transverse Ranges, CA. *Geol. Soc. Am. Bull.*, **100**, 500–13.

Rockwell, T.K., Keller, E.A., Clark, M.N., and Johnson, D.L. (1984). Chronology and rates of faulting of Ventura River terraces. *Geol. Soc. Am. Bull.*, **95**, 1466–74.

Royden, L.H. (1988). Late Cenozoic tectonics of the Pannonian Basin system. *Am. Assoc. Petrol. Geol. Mem.*, **45**, 27–48.

Royden, L.H. and Báldi, T. (1988). Early Cenozoic tectonics and paleography of the Pannonian and surrounding regions. *Am. Assoc. Petrol. Geol. Mem.*, **45**, 1–16.

Royden, L.H. and Horváth, F. (1988a). Introduction to the Pannonian region. In L.H. Royden and F. Horváth (eds.) *The Pannonian Basin, a study in basin evolution*, American Association of Petroleum Geologists, Memoir 45, 394 pp.

Rumpler, J. and Horváth, F. (1988b). Some representative seismic reflection lines from the Pannonian basin and their structural interpretation. *Am. Assoc. Petrol. Geol. Mem.*, **45**, 153–69.

Ruppel, E. T., O'Neil, J. M., and Lopez, D. A. (1983). *Preliminary geologic map of the Dillon 1° × 2° quadrangle, Montana*. U.S. Geol. Survey Open-File Report 83–168.

Russ, D. R. (1982). *Style and significance of surface deformation in the vicinity of New Madrid, MO*. U.S. Geol. Survey Prof. Paper 1236–H, pp. 95–114.

Russell, R. J. (1939). Louisiana stream patterns. *Am. Assoc. Petrol. Geol.*, **23**, 1199–227.

Russell, R. J. (1954). Alluvial morphology of Anatolian rivers. *Assoc. Am. Geog. Ann.*, **44**, 363–91.

Rust, B. R. and Koster, E. H. (1984). Coarse alluvial deposits. In Walker, R. G. (ed.) *Facies models*. Geoscience Canada Reprint Series 1, pp. 53–69.

Rust, B. R. and Legun, A. S. (1983). *Modern anastomosing-fluvial deposits in arid central Australia, and a Carboniferous analogue in New Brunswick, Canada*. International Association of Sedimentologists Special Publ. 6, pp. 385–92.

Said, R. (1981). *The geological evolution of the River Nile*. Springer-Verlag, NY.

Said, R. (ed.) (1990). *The geology of Egypt*. Balkema, Rotterdam.

Sanford, A. R., Mott, R. P., Jr, Shuleski, P. J., Rinehart, E. J., Caravalle, F. J., Ward, R. M., and Wallace, T. C. (1977). *Geophysical evidence for a magma body in the crust in the vicinity of Socorro, New Mexico*. Am. Geophy. Union, Geophysical Monograph 20, pp. 385–404.

Sanz, V. P. (1974). Geologia preliminar del area Tigre-Corrientes en el Nororiente peruano. *Boletin de la Sociedad Geologica del Peru*, **44**, 106–27.

Saucier, R. T. (1964). *Geological investigation of the St. Francis basin*. U.S. Army Engineers, Waterways Experiment Station, Vicksburg, MS, Tech. Report 3–649, 95 pp.

Saucier, R. T. (1977). *Effects of the New Madrid Earthquake series in the Mississippi alluvial valley*. Miscellaneous Paper S-77-5, Soils and Pavements Laboratory, U.S. Army Engineers Waterways Experiment Station, Vicksburg, MS, 10 pp.

Saucier, R. T. (1987). *Geomorphological interpretations of late Quaternary terraces in western Tennessee and their regional tectonic implications*. U.S. Geol. Survey Prof. Paper 1336–A, 19 pp.

Saucier, R. T. (1989). Evidence for episodic sand-blow activity during the 1911–1912 New Madrid (Missouri) earthquake series. *Geology*, **17**, 103–6.

Saucier, R. T. (1991). Geomorphology, stratigraphy, and chronology. In Morrison, R. B. (ed.) *Quaternary nonglacial geology: the Geology of North America*, vol. K-2, pp. 550–63.

Saucier, R. T. and Fleetwood, A. R. (1970). Origin and chronological significance of late Quaternary terraces, Ouachita River, Arkansas and Louisiana. *Geol. Soc. Am. Bull.*, **81**, 869–90.

Schumm, S. A. (1963). *Disparity between present rates of denudation and orogeny*. U.S. Geol. Survey Prof. Paper 454–H, pp. H1–H13.

Schumm, S. A. (1968). Speculations concerning paleohydrologic controls of terrestrial sedimentation. *Geol. Soc. Am. Bull.*, **79**, 1573–88.

Schumm, S. A. (1972). Fluvial Paleochannels. In Rigby, J. K. and Hamblin, W. K. (eds.) *Recognition of ancient sedimentary environments*. Society of Sedimentary Geologists Special Publ. 16, pp. 98–107.

Schumm, S. A. (1977). *The fluvial system*. John Wiley and Sons, New York, 338 pp.

Schumm, S. A. (1979). Geomorphic thresholds: the concept and its applications. *Inst. Br. Geog. Trans.*, **4**, 485–515.

Schumm, S. A. (1981). *Evolution and response of the fluvial system: sedimentologic implications*. Society of Economic Paleontologists and Mineralogists Spec. Publ. 31, pp. 19–29.

Schumm, S.A. (1986). *Alluvial river response to active tectonics, studies in geophysics, active tectonics*. National Academy Press, Washington D.C., pp. 80–94.

Schumm, S.A. (1993). River response to baselevel change: implications for sequence stratigraphy. *J. Geology*, **101**, 279–94.

Schumm, S.A. and Ethridge, F.G. (1994). Origin, evolution, and morphology of fluvial valleys. In Dalrymple, R.W., Boyd, R., and Zaitlin, B.A. (eds.) *Incised-valley systems: origin and sedimentary sequences*. Society Sedimentary Geologists Special Publ. 51, pp. 11–27.

Schumm, S.A. and Galay, V.J. (1994). The River Nile in Egypt. In Schumm, S.A. and Winkley, B.R. (eds.) *The variability of large alluvial rivers*. Am. Soc. Civil Engineers Press, NY, pp. 75–100.

Schumm, S.A. and Harvey, M.D. (1985). Preliminary geomorphic evaluation of the Sacramento River (Red Bluff to Butte Basin). Unpublished report to Sacramento District, Corps of Engineers, 57 pp.

Schumm, S.A. and Khan, H.R. (1972). Experimental study of channel patterns. *Geol. Soc. Am. Bull.*, **83**, 1755–70.

Schumm, S.A. and Rea, D.K. (1995). Sediment yield from disturbed earth systems. *Geology*, **23**, 391–7.

Schumm, S.A. and Spitz, W.J. (1997). Geological influences on the lower Mississippi River and its alluvial valley. *Eng. Geol.*, **45**, 245–61.

Schumm, S.A., Gregory, D.I., Lattman, L.H., and Watson, C.C. (1984). *River response to active tectonics: Final report to National Science Foundation*. Water Engineering & Technology, Inc., Fort Collins, CO, 125 pp.

Schumm, S.A, Mosley, M.P., and Weaver, W.E. (1987). *Experimental fluvial geomorphology*. John Wiley, NY, 413 pp.

Schumm, S.A., Rutherfurd, I.D., and Brooks, J. (1994). Pre-cutoff morphology of the lower Mississippi River. In Schumm, S.A. and Winkley, B.R. (eds.) *The variability of large alluvial rivers*. Am. Soc. Civil Eng. Press, NY, pp. 13–44.

Schumm, S.A., Erskine, W.D., and Tilleard, J.W. (1996). Morphology, hydrology, and evolution of the anastomosing Ovens and King Rivers, Victoria, Australia. *Geol. Soc. Am. Bull.*, **108**, 1212–24.

Schwartz, R. (1982). Broken Early Cretaceous foreland basin in southwestern Montana: Sedimentation related to tectonism. In Powers, R.B. (ed.) *Geologic studies of the Cordilleran thrust belt*. Rocky Mountain Association of Geologists, pp. 159–83.

Scott, J.C. (1972). Geology of Monroe County: Geol. Survey Alabama, Special Map 101.

Scott, A.J. and Horne, J.C. (1989). Depositional models, sequence stratigraphy and tectonic controls on valley-fill sequences [abst]. *CSPG Reservoir*, **16**(8), 1.

Seeber, L. and Gornitz, V. (1983). River profiles along the Himalayan arc as indicators of active tectonics. *Tectonophysics*, **92**, 335–67.

Servant, M., Fontes, J.C., Rieu, M. and Saliège, J.F. (1981). Phases climatiques arides Holocène dans le sud-ouest de l'Amazonie. *Comptes Rendus de l'Académie des Sciences de Paris*, **292**, 1295–1297.

Shanley, K.W. and McCabe, P.J. (1991). Predicting facies architecture through sequence stratigraphy – an example from the Kaiparowits Plateau, UT. *Geology*, **19**, 742–5.

Shanley, K.W. and McCabe, P.J. (1994). Perspectives on the sequence stratigraphy of continental strata. *Am. Assoc. Petrol. Geol. Bull.*, **78**(4), 544–68.

Shanley, K.W., McCabe. P.J. and Hettinger, R.D. (1992). Tidal influences in Cretaceous fluvial strata from Utah, USA: a key to sequence stratigraphy of continental strata. *Sedimentology*, **39**, 905–30.

Shuster, M.W. and Steidtman, J.R. (1987). Fluvial-sandstone architecture and thrust-induced subsidence, northern Green River Basin, Wyoming. In Ethridge, F.G., Flores, R.M., and Harvey, M.D. (eds.) *Recent developments in fluvial sedimentology*. Society of Economic Paleontologists and Mineralogists, Special Publ. 39, pp. 279–85.

Sieh, K.E. and Jahns, R.H. (1984). Holocene activity of the San Andreas fault at Wallace Creek, CA. *Geol. Soc. Am. Bull.*, **95**, 883–96.

Slack, P.B. (1981). Paleotectonics and hydrocarbon accumulation, Powder River Basin, Wyoming. *Am. Assoc. Petrol. Geol. Bull.*, **65**, 730–43.

Smith, D.G. (1983). *Anastomosed fluvial deposits: modern examples from western Canada*. International Association of Sedimentologists Special Publ. 6, pp. 155–68.

Smith, D.G. and Smith, N.D. (1980). Sedimentation in anastomosed river systems: examples from alluvial valleys near Banff, Alberta. *J. Sed. Petrol.*, **50**, 157–64.

Smith, N.D. and Pérez-Arlucea (1994). Fine-grained splay deposition in the avulsion belt of the lower Saskatchewan River, Canada. *J. Sed. Res.*, **B64**(2), 159–68.

Smith, N.D., Cross, T.A., Dufficy, J.P., and Clough, S.R. (1989). Anatomy of an avulsion. *Sedimentology*, **36**, 1–23.

Smith, R.B. and Sbar, M.L. (1974). Contemporary tectonics and seismicity of the western United States with emphasis of the Intermountain Seismic Belt. *Geol. Soc. Am. Bull.*, **85**, 1205–18.

Sparling, D.R. (1967). Anomalous drainage pattern and crustal tilting in Ottawa County, Ohio. *Ohio J. Sci.*, **67**, 378–81.

Spitz, W.J. and Schumm, S.A. (1997). Tectonic geomorphology of the Mississippi valley between Osceola, Arkansas and Friars Point, Mississippi. *Eng. Geol.*, **46**, 259–80.

Srivastava, P., Parkash, B., Sehgal, J.L. and Kumar, S. (1994). Role of neotectonics and climate in development of the Holocene geomorphology and soils of the Gangetic Plains between the Ramganga and Rapti rivers. *Sed. Geol.*, **94**, 129–51.

State of California, Department of Water Resources (1984). *River Atlas Middle Sacramento River spawning gravel study*, 34 pp.

Steel, R.J., Maehle, S., Nilsen, H., Roe, S.L., and Spinnanger, A. (1977). Coarsening-upward cycles in the alluvium of Hornelen Basin (Devonian), Norway: sedimentary response to tectonic events. *Geol. Soc. Am. Bull.*, **88**, 1124–34.

Stein, R.S. and King, G.C.P. (1984). Seismic potential revealed by surface folding, Coalinga, CA, earthquake. *Science*, **224**, 867–72.

Sternberg, H.O. (1950). Vales tectonicas na planicie amazonicas? *Revista Brasilera de Geografia*, **4**(12), 511–34.

Sternberg, H.O. (1955). Sismicite et morphologie en Amazonie bresilienne. *Annales de Geographie*, **342**, 97–105.

Stevens, G.R. (1974). *Rugged landscape, the geology of central New Zealand*. A.H. and A.W. Reed, Wellington, 286 pp.

Stewart, J.H. (1978). Basin Range structure in western North America, a review. *Geol. Soc. Am. Mem.*, **152**, 1–31.

Stickney, M.C. and Bartholomew, M.J. (1987). Seismicity and late Quaternary faulting of the northern Basin and Range Province, Montana and Idaho. *Seismol. Soc. Am. Bull.*, **77**, 1602–25.

Stiglish, G. (1904). *Ultimas exploraciones ordenadas por la Junta de vias fluviales a los rios Ucayali.* Madre de Dios, Paucartambo, y Urubamba, Oficina Tipografica de "La Opinion Nacional."

Sykes, L. R. (1978). Intraplate seismicity, reactivation of preexisting zones of weakness, alkaline magmatism, and other tectonism postdating continental fragmentation. *Rev. Geophys. Space Phys.*, **16**(4), 621–88.

Sztanó, O. and Tari, G. (1993). Early Miocene basin evolution in northern Hungary, tectonics and eustacy. *Tectonophysics*, **226**, 485–502.

Talbot, C. J. (1977). Inclined and asymmetric upward-moving gravity structures. *Tectonophysics*, **42**, 159–81.

Tari, G., Baldi, T., and Baldi-Beke, M. (1993). Paleogene retroarc flexural basin beneath the Neogene Pannonian Basin: a geodynamic model. *Tectonophysics*, **226**, 433–55.

Tator, B. A. (1958). The aerial photograph and applied geomorphology. *Photogram. Eng.*, **24**, 549–61.

Thomas, G. (1993). Basement paleotopography key in Colorado Marrow channel oil play. *Oil Gas J.*, July 19, 56–62.

Titheridge, D. G. (1993). The influence of half-graben syn-depositional tilting on thickness variation and seam splitting in the Brunner Coal Measures, New Zealand. *Sed. Geol.*, **87**, 195–213.

Todd, S. P. and Went, D. J. (1991). Lateral migration of sand-bed rivers: examples from the Devonian Glashabeg Formation, SW Ireland and the Cambrian Alderney Sandstone Formation, Channel Islands. *Sedimentology*, **38**, 997–1020.

Tolman, C. F. and Poland, J. F. (1940). Ground-water, salt-water infiltration, and ground-surface recession in Santa Clara Valley, Santa Clara County, CA. *Am. Geophy. Union Trans.*, **21**, 23–35.

Törnqvist, T. (1994). Middle and late Holocene avulsion history of the River Rhine (Rhine-Meuse delta, Netherlands). *Geology*, **22**, 711–14.

Tricart, J. (1974). *Structural geomorphology*. Longmans, NY, 305 pp.

Tricart, J. (1977). Types de lits fluviaux en Amazonie bresilienne. *Ann. Geog.*, **473**, 1–51.

Twidale, C. R. (1966). Late Cenozoic activity of the Selwyn upwarp, northwest Queensland. *J. Geol. Soc. S. Austral.*, **13**, 491–4.

Twidale, C. R. (1971). *Structural landforms*. Australian National Univ. Press, Canberra.

U.S. Congress (1911). *Examination and survey of Sacramento River*. 62 Congress, 1st Session House Documents, 76 pp.

U.S. Department of Commerce (1972). *News*, Friday September 22, 1972, NOAA 72–122, 3 pp.

van Gelder, A., van den Berg, J. H., Guodong, C., and Chunting, X. (1994). Overbank and channel fill deposits of the modern Yellow River delta. *Sed. Geol.*, **90**, 293–305.

Vanicek, P. and Nagy, D. (1980). The map of contemporary vertical crustal movements in Canada. *EOS*, **61**, 145–7.

Veatch, A. C. (1906). *Geology and underground water resources of northern Louisiana and southern Arkansas*. U.S. Geol. Survey Prof. Paper 46, 442 pp.

Villarejo, A. (1988). *Asi es la Selva*. Centro de Estudio Teologicos de la Amazonia, Iquitos, 330 pp.

Volkov, N.G., Sokolovski, I.L., and Subbotin, A.I. (1967). *Effect of recent crustal movement on the shape of longitudinal profiles and water levels in rivers*. International Association of Scientific Hydrology Publ. 75, pp. 105–16.

Von Bandat, H.F. (1962). *Aerogeology*. Gulf, Houston, TX, 350 pp.

Wallace, R.E. (1967). Notes on stream channels offset by the San Andreas Fault, southern coast ranges California. *Stanford Univ. Pub. (Geol. Sci.)*, **11**, 6–20.

Wallace, R.E. (1975). *The San Andreas fault in the Carrizo plain – Tremblor Range region, CA*. Calif. Div. Mines and Geology Special Report 118, pp. 241–50.

Wallace, R.E. (1985). *Active tectonics – impact on society, overview and recommendations*. National Academy of Science, Washington, D.C.

Walters, W.H., Jr. (1975). Regime changes of the lower Mississippi River. Unpublished M.S. thesis, Colorado State University, 129 pp.

Walters, W.H., Jr. and Simons, D.B. (1984). Long-term changes of lower Mississippi River meander geometry. In Elliott, C.M. (ed.) *River Meandering*. American Society of Civil Engineers, New York, pp. 318–29.

Wang, K.K. (1952). Geology of Ouachita Parish. *Louisiana Geol. Surv. Bull.*, **28**, Baton Rouge, LA, 126 pp.

Weimer, R.J. (1984). Relationship of unconformities, tectonics and sea level changes, Cretaceous Western Interior, United States of America. In Schlee, J.S. (ed.), *Interregional unconformities and hydrocarbon accumulation*. American Association of Petroleum Geologists, Memoir **36**, 7–36.

Weimer, R.J., Emme, J.J., Farmer, C.L., Anna, L.O., Davis, T.L., and Kidney, R.L. (1982). Tectonic influence on sedimentation, Early Cretaceous, east flank Powder River Basin, Wyoming and South Dakota. *Colorado Sch. Mines Quart.*, **77**(4), 1–61.

Weimer, R.J., Sonnenberg, S.A., and Young, G.B. (1986). Wattenberg field, Denver Basin, Colorado. In C.W. Spencer and R.F. Mast (eds.) *Geology of tight gas reservoirs*. American Association of Petroleum Geologists, Studies in Geology, **24**, 143–64.

Welch, D.M. (1973). Channel form and bank erosion, Red River, Manitoba. In *Fluvial Processes and Sedimentation*. Proceedings of a Hydrology Symposium, University of Alberta, pp. 284–93.

Wells, N.A. and Dorr, J.A. (1987). Shifting of the Kosi River, northern India. *Geology*, **15**, 204–18.

Wescott, W.A. (1993). Geomorphic thresholds and complex response of fluvial systems – some implications for sequence stratigraphy. *Am. Assoc. Petrol. Geol. Bull.*, **77**, 1208–18.

Willemin, J.H. and Knuepfer, P.L.K. (1994). Kinematics of arc-continent collision in the eastern central range of Taiwan inferred from geomorphic analysis. *J. Geophys. Res.*, **99**, 20 267–80.

Willis, B.J. and Behrensmeyer, A.K. (1994). Architecture of Miocene overbank deposits in northern Pakistan. *J. Sed. Res.*, **B64**(1), 60–67.

Wise, D.U., Belt, E.S., and Lyons, P.C. (1991). Clastic diversion by fold salients and blind thrust ridges in coal-swamp development. *Geology*, **19**, 514–17.

Wohl, E.E. and Georgiandi, A.G. (1994). Holocene paleomeanders along the Sejm River, Russia. *Zeit. Geomorphol.*, **38**, 299–309.

Wollard, G.P. (1958). Areas of tectonic activity in the United States as indicated by earthquake epicenters. *Am. Geophys. Union Trans.*, **39**, 1139–50.

Wood, D. H. and Guevara, E. H. (1981). *Regional structural cross sections and general stratigraphy, East Texas Basin.* University of Texas, Bureau Economic Geology, 21 pp.

Woolfe, K. J. (1992). *Complex sand sheets from non-avulsive meandering streams.* Fifth meeting of the Australia and New Zealand Geomorphology Research Group, Abstract, Volume 7.

Yeats, R. S., Sieh, K., and Allen, C. R. (1997). *The geology of earthquakes.* Oxford University Press, NY, 568 pp.

Yerkes, R. F. and Castle, R. O. (1969). *Surface deformation associated with oil and gas field operations in the United States.* International Association of Scientific Hydrology Publ. 88, pp. 55–66.

Yeromenko, V. Y. and Ivanov, V. P. (1977). Study of river meandering in the search for tectonic structures (as exemplified by the middle Dniester Region). *Sov. Hydrol.,* **16**, 9–14.

Youssef, M. I. (1968). Structural pattern of Egypt and its interpretation. *Am. Assoc. Petrol. Geol. Bull.,* **52**, 601–14.

Yoxall, W. H. (1969). The relationship between falling baselevel and lateral erosion in experimental streams. *Geol. Soc. Am. Bull.,* **80**, 1379–84.

Zhou, Z. and Pan, X. (1994). Lower Yellow River. In Schumm, S. A. and Winkley, B. R. (eds.) *The Variability of Large Alluvial Rivers.* American Society of Civil Engineers Press, New York, pp. 363–94.

Zoetemeijer, R., Cloetingh, S., Sassi, W., and Roure, F. (1993). Modeling of piggyback-basin stratigraphy; record of tectonic evolution. *Tectonophysics,* **226**, 253–69.

Zuchiewicz, W. (1979). A possibility of application of the theoretical longitudinal river's profile analysis to investigations of young tectonic movements. *Ann. Natl Soc. Geol. Pologne,* **49**, 327–42.

Zuchiewicz, W. (1980). Young tectonic movements and morphology of the Piening Mountains (Polish western carpathians). *Ann. Natl Soc. Geol. Pologne,* **50**, 263–300.

Index